C0-ARS-830

Laboratory Experiments in Environmental Chemistry

Laboratory Experiments in Environmental Chemistry

D. Neal Boehnke, Ph. D.

Department of Chemistry, Jacksonville University

R. Del Delumyea, Ph. D.

Former Director, Millar Wilson Laboratory for Chemical Research

PRENTICE HALL
UPPER SADDLE RIVER, NEW JERSEY 07458

Library of Congress Cataloging-in-Publication Data

Boehnke, D. Neal,
Laboratory experiments in environmental chemistry / D. Neal Boehnke and R. Del Delumyea
p. cm.
Includes bibliographical references and index.
ISBN 0-13-917171-1 (pbk.)
1. Environmental chemistry Laboratory manuals. I. Delumyea, R. Del, . II. Title
TD193.B64 2000
628.5'028—dc21 99-24941
CIP

Editor-in-Chief: Paul F. Corey
Assistant Vice President of Production and Manufacturing: David W. Riccardi
Executive Managing Editor: Kathleen Schiaparelli
Assistant Managing Editor: Lisa Kinne
Production Editor: Linda DeLorenzo
Marketing Manager: Steven Sartori
Editorial Assistant: Betsy Williams
Manufacturing Manager: Trudy Pisciotti
Art Director: Jayne Conte
Cover Designer: Bruce Kenselaar

©2000 by Prentice-Hall, Inc.
Upper Saddle River, New Jersey 07458

All rights reserved. No part of this book may be reproduced, in any form or by any means, without permission in writing from the publisher.

Printed in the United States of America

10 9 8 7 6 5 4 3 2

ISBN 0-13-917171-1

Prentice-Hall International (UK) Limited, *London*
Prentice-Hall of Australia Pty.Limited, *Sydney*
Prentice-Hall Canada Inc., *Toronto*
Prentice-Hall Hispanoamericana, S.A., *Mexico*
Prentice-Hall of India Private Limited, *New Delhi*
Prentice-Hall of Japan, Inc., *Tokyo*
Prentice-Hall (*Singapore*) Pte Ltd
Editora Prentice-Hall do Brasil, Ltda., *Rio de Janeiro*

Dedication

This book is dedicated to John Austin, Larry Gilmore, Bob Harris, and William E. Sweisgood. Their untiring efforts and dedication to preserve the environment and its inhabitants have succeeded in raising the level of the public's awareness of many environmental problems. Their works have led to changes in policies and laws resulting in profound long-term effects on environmental quality and protection.

CONTENTS

* This experiment requires more background than general chemistry.

* This experiment requires more background than general chemistry.

PREFACE

This laboratory manual is the result of a search for one. Environmental chemistry is a relatively new discipline, and texts, especially laboratory manuals, in this area are few. This manual is an outgrowth of an analytical chemistry laboratory manual that one of the authors wrote more than 20 years ago. To stimulate student interest, standard unknowns were frequently replaced by soil, sediment, and water samples taken from the local environment, which proved to have a wide variety of pollutants.

Since the interest in environmental chemistry crosses many disciplines, it is likely that a significant number of students taking this course might have only a background in general chemistry. To make this laboratory course feasible and functional for students with varied backgrounds, a review of basic laboratory practices is provided in the Introduction to Laboratory Work. Each experiment also begins with fundamental concepts and basic theory, which are then developed into the more advanced concepts. In essence, this approach fills in important background material for less experienced students and provides a valuable review for the more experienced.

The often long introductory sections preceding each experiment are included for several reasons. Since the average student in this course may have a limited background in chemistry, these discussions are included to assist the students in understanding and discussing their results. Since there are no "unknowns" in this course, the students' results must be understood in order to be discussed. The authors believe that "understanding" results is one of the more important goals a student can achieve in this course.

Second, to decide where samples should be collected it is important to understand the origin, nature, and fate of environmental pollutants, topics covered in the introductory discussions. One should not attempt to collect hydrocarbons, for example, from a pond on a farm far away from any source of petroleum-derived products. Yet the pond may be an excellent source of samples for studying potassium, fluoride, or manganese levels.

The authors have extensive experience in teaching analytical chemistry as well as practical experience. Both have carried out, and directed, extensive environmental projects. When we began a course in environmental chemistry, we used our experience to put together a list of experiments, which eventually evolved into this manual. Most of the experiments presented here have been used for six semesters at two universities. Many of the experiments have been refined by a critical examination by the instructors and from students' input. The students initially doing the experiments were patient with the unorganized efforts of the authors, and their suggestions resulted in a data-treatment section that greatly helped to organize and interpret experimental results.

This manual begins with a review of basic statistical concepts, with applications using basic chemical equipment that will be used throughout this course. This is followed by an experiment on sample collection and preservation, a topic that reappears throughout most of the experiments. Following are two broad types of experiment: Part III, emphasizing properties of inorganic substances, and Part IV, covering the properties of organic matter. These two areas include the determination of physical characteristics, carrying out instrumental methods of analysis, and using standard methods adopted by state and federal agencies.

Experiments become increasingly complex when flame-ionization gas chromatography and atomic absorption are introduced. We have included two experiments centered around gas chromatography and two on atomic absorption spectrometry. These are important methods in environmental chemistry, even at the elementary level.

In addition to analytical-type experiments we have included experiments that focus on inorganic, organic, and physical chemistry. The middle experiments, comprising about six to seven in all, cover elementary instrumental methods, including spectroscopy, and gas chromatography.

The later experiments cover two areas, the first involving physical chemistry: surface tension and kinetics. Environmental chemistry is a very broad field and encompasses all areas of chemistry, and we have tried to show this by including a wide variety of experiments. Although this may seem out of place in an environmental manual at any level, environmental chemistry is interdisciplinary, covering not only the chemical fields of analytical chemistry, instrumental analysis, organic, inorganic and physical chemistry, but also reaching into the biological (biochemistry) and mathematical (statistics) realms.

Finally we end with three experiments involving the atmosphere, which the authors believe to be unusual in a manual at this level and, at any rate, are not usually found in any manual for educational purposes.

The experiments are arranged in a logical, pedagogical order, the way a typical environmental problem is approached.

Many of the experiments can be carried out using the CBL (calculator-based laboratory) program (details are given in the Instructor's Manual). This includes pH, conductivity, and colorimetry measurements. Temperature and barometric pressure can also be measured with these units. This allows computer-treatment of results without directly interfacing to a computer, which would require many computers. Using a T.I. 85 and a CBL unit, data are acquired and subsequently down-loaded to a computer for data and graphical analysis. A major advantage of using a CBL is its portability. The student is also encouraged to use a computer to generate laboratory reports.

The authors have students use Graphical Analysis for Windows(Vernier Software) for generating data tables and graphs and for carrying out linear regressions. Students are encouraged to use their own computer expertise (many have a lot) to treat data.

Several of the experiments found in this manual can be found in other manuals (in different form), but the emphasis and techniques used here are often very different, a result of dealing with real samples from the environment. An analysis for manganese or fluoride must take into account interferences and other environmental factors (such as complexation).

The experiments presented in this manual are a result of the authors' judgment about what is important for the student beginning environmental studies in chemistry. A number of experiments are simply "analyses" where the student measures "hardness," manganese or fluoride levels, etc.; but still the student must explain what the results mean.

Many students of science learn or do science because at one time they were excited by experimental observations. Experiment, ultimately, is the scientist's reality and presents the phenomena that are ultimately to be understood. It is unfortunate that students in their experimental studies are often not able to follow their natural inclinations to explore the unknown. Environmental chemistry provides an arena for overcoming such limitations.

The authors have had significant and important input from the following faculty at Jacksonville University: Drs. G. Edwin Lewis, Fred Senftleber, and Lucy Sonnenberg. More than 100 students have used most of this manual, and we are greatly indebted to them for their patience and constructive comments.

Ms. Evelyn Sparrow was very important in the early typing and critical comments on this work. Her work went far beyond the routine and resulted in style in the final product. Ms. Judy SanSocie made many useful suggestions and helped in the technical aspects of word processing. Cheryl Boehnke was critical in the completion of this work, including typing, criticisms, figures, and important additions of scientific content. Mark Boehnke was helpful in sharing with us his computer expertise in many of the mechanical aspects in the completion of this manual. David Boehnke was helpful in several aspects in the production of this work.

The authors gratefully acknowledge the comments of the reviewers. Their expertise and extensive experience, which they shared with us, made a great difference in the final product. We therefore thank the following: Bob Allen, Arkansas Tech University; Sharelle Campbell, Louisiana State University in Shreveport; Barry A. Coddens, Northwestern University; Stephen R. Daniel, Colorado School of Mines; Annmarie Eldering, University of Iowa; R. Max Ferguson, Eastern Connecticut State University; Barry R. Ganong, Mansfield University; Barbara Kebbekus, New Jersey Institute of Technology; William Lammela, Nazareth College; Kathleen E. Murphy, Daemen College; Carol Sands, University of Wisconsin–Platteville; Jeffery A. Schneider, State University of New York–Oswego; Stephen Steppenuck, Keene State College; William C. Trogler, University of California, San Diego; Robert C. Troy, Central Connecticut State University.

Finally, we wish to acknowledge the patient, understanding, and very competent help from the editors and production staff of Pearson Education: Mr. John Challice, Mr. Matt Hart, Ms. Betsy Williams, Ms. Connie Wu, and Ms. Linda DeLorenzo.

Finally, the authors acknowledge their responsibility for the final product and would appreciate comments for the improvement of this manual.

D. Neal Boehnke

R. Del Delumyea

To the Student

We now discuss several important points relative to correct scientific procedure, safety, and successfully carrying out the experiments in this manual. As in any laboratory course in science, a laboratory notebook must be kept. The specifics as to deletions, what to include, and other details depend on the individual instructor and will not be discussed here. However, there are basic rules for any laboratory notebook: use a hard-bound book, make all entries in ink, use a title at the beginning of each experiment, and put the date and your initials on each page you use. This is the minimum needed, but your instructor may require a lot more. It is obvious that your name and address should be on the front of the book. It is also desirable to make carbon copies of your data and to keep the copies in a place separate from your lab book.

Because the Introduction and Theory sections of the experiments may be very lengthy, at least read (and take notes in your lab book, if you like) the Objective(s), and the basic theory relating to the experiment, and most importantly, read the Experimental Procedure before beginning the experiment. If you have questions about the procedure, the best time to ask them is before you start the experiment (and make mistakes).

Safety is the foremost consideration in any laboratory course, but in environmental chemistry it is of even greater importance since there are several new and subtle dangers to be faced. To be clear, we enumerate these factors:

(1) Safety glasses must be worn at <u>all</u> times when you are in the lab and when collecting and treating samples in the field.

(2) A wide variety of chemical reagents are used throughout this course. Use gloves when handling dangerous chemicals. When pouring from most bottles liquid will dribble down the sides, and it should be assumed that when you touch a reagent bottle it will have liquid on its side. It is good practice to use wet paper towels to wipe a bottle before handling it.

(3) There are some chemicals that cannot simply be washed off if physical contact is made. Sodium hydroxide and chlorinated carboxylic acids penetrate the skin and cannot be simply washed off. Rinsing with water for several minutes is required, or a compress of sodium bicarbonate is helpful if rinsing does not suffice.

(4) If a reagent has any significant volatility (that is, you can smell it), use it in a fume hood (with the hood on). Many chemicals are poisonous to the extent that if you smell them, you are getting a toxic dose. Fume hoods are one factor responsible for increasing the life span of chemists.

(5) The instructor must be present during environmental sample collection, not only for reasons of safety, but also to teach proper sampling procedures and also to point out factors that may affect results (for example, the presence of a nearby outfall or smoke plume). Also, when collecting samples near water, it is imperative that a life vest be worn.

When collecting samples from a boat, it is absolutely essential to have an experienced captain. It is often necessary to take samples from very shallow water and at other times from areas where close quarters are encountered. Also, it may be desirable to collect samples in the rain or at night, and experience in handling the boat is absolutely necessary.

In addition to an experienced captain, an experienced student aide is very important in carrying out various duties, including anchoring, navigation, and proper positioning of the boat for sample collection, and sometimes helping to collect samples or showing students how to use sampling apparatus.

(6) A less obvious safety factor in environmental studies is the biological content of the samples. Especially in warmer climates samples will accommodate a large number of a wide variety of bacteria (and viruses). If samples are kept in a warm lab for a period of time an exponential rate of growth is possible and may result in a high level of disease-causing microorganisms. Thus, if samples are stored before analysis, they should be stored at about 4°C and preferably preserved with acid. Even when stored and preserved properly, samples must be handled with gloves.

When collecting samples, it is important to be constantly vigilant. Do not collect samples from a sewage outfall for obvious reasons. If a sample smells like sewage, it may be wise to not use it.

You are encouraged to explore your local environment. Determine what activities cause local environmental problems, find local outfalls and storm drains, etc. By learning what activities may cause pollution problems, it is possible to know where to sample and what to sample for.

Laboratory Experiments in Environmental Chemistry

Fundamentals for Studies of Environmental Chemistry

1

Introduction to Laboratory Work

Most students taking environmental chemistry have had a minimum of two terms of introductory college chemistry, with laboratory. Most of the experiments in this manual are directed at this level, but some of the later experiments are better suited to students with more background, in particular analytical chemistry, instrumental analysis, or organic chemistry. Experience has shown, however, that a good student with a general chemistry background can perform all the experiments in this manual. More advanced experiments are indicated by an asterisk in the Table of Contents.

Most schools offering a major in environmental science and an environmental chemistry course with laboratory are probably equipped to perform the majority of the experiments in this manual.

Most students above the freshman level have practical knowledge of basic laboratory glassware, including volumetric flasks, pipets, and burets, and have probably done several titrations. However, these students may not appreciate finer points, such as the meaning of Class A glassware, uncertainties in estimating volumes, and tolerances. Also, many students in this course may not have used an analytical balance. To review, and in some cases, preview, these areas, a brief introduction is provided on the use and care of the analytical balance, burets, pipets, and volumetric flasks.

One of the more difficult topics to consider in this course is waste management and disposal. Much has changed in this area in recent years. Practices such as evaporating solvents and wastes in fume hoods are no longer acceptable.

Concern for the environment and strict environmental regulations, with potential high penalties for noncompliance, have resulted in waste minimization and control programs in government, industry, and academic laboratories. These programs encourage using smaller quantities of chemicals, replacing toxic chemicals with less toxic ones, and segregating, characterizing, and storing used chemicals for proper disposal by qualified contractors.

Some reviewers of this manual indicated that no wastes can be rinsed down lab drains in their city or state. Others indicated that quantity and type of waste are determining factors for drain disposal. Most schools have safety officers tasked to develop, implement, and manage plans to deal with hazardous wastes for the entire institution. Environmental regulations require the tracking of wastes from their generation and storage to their ultimate disposal.

Environmental regulations vary from state to state, and even federal regulations, such as those promulgated by the U. S. Environmental Protection Agency (EPA), may differ in interpretation between the various EPA regions of the United States. In addition, city and other local ordinances must be followed.

Because users of this manual will have different guidelines for waste disposal, only general instructions are provided which closely follow the regulations pertaining to disposal in the authors' area. Users of this manual should follow the regulations for their particular locations and are advised to consult with their institutional hazardous waste coordinator(s).

I. Safety Concerns

In chemistry labs the most important safety problems are cuts, burns, and chemical contact (skin or eyes). In the environmental chemistry course there are additional safety concerns, some associated with the nature of the samples and others associated with the sampling process itself.

Chemical Safety glasses must be worn at all times in the chemistry laboratory to prevent injury to the eyes. For particularly dangerous operations, safety goggles and a face shield may also be required.

Wearing gloves can prevent chemical burns on the hands, which frequently occurs because of liquid dripping down the sides of a bottle. A disadvantage of some gloves is that they are slippery and this may cause the bottle being handled to drop and break. This can be avoided by using textured-grip, powder-free gloves. Some important glove types and their properties are listed in the Table I-1.

Table I-1 Safety Gloves and Their Characteristics

Type of Glove	Glove Characteristics
Latex	Low in particulates and resistant to most laboratory chemicals
Polyethylene	Cheap, generally liquid-proof, and chemically resistant
PVC vinyl	Resists common chemicals and oils

Gloves are safety rated on the basis of three factors: their degradation, chemical breakthrough time, and permeation rate. If gloves need to be used over an extended period of time, they should be changed periodically since most gloves will ultimately permit the migration of chemicals through to the skin. Generally powder-free gloves are preferable to avoid contamination and slipperiness. Finger cots are useful for handling small items such as weighing boats or weighing bottles.

Biological A distinguishing characteristic of environmental samples is their possibility of being biologically contaminated. One of the authors once collected samples of what was supposed to be leachate from a hazardous waste site. The appearance and odor of the samples led to their identification as runoff from an improperly installed septic tank. If sediment samples are stored at room temperature for an extended period, there is a possibility that bacteria present will multiply to a great extent and pose a health hazard. In urban areas where there are sewage outfalls, there is also the possibility of large populations of bacteria, even from treated sewage.

When collecting samples of sediment or water, gloves should be worn and all possible contact with the skin avoided. When stored, these samples should be kept at about 4°C to minimize bacterial growth. When weighing out these samples, gloves should be worn and any spillage wiped up immediately and disinfected with a disinfectant detergent.

Sample Collection Close supervision is important when collecting samples from the environment. The instructor can not only advise about procedures, but can also point out local features that may be important in discussing and interpreting results. Some areas of a stream might be impacted by sewage outfalls, while other areas may be subjected to industrial effluents. Stormwater drains must also be considered in interpreting results if they are in the sampling area.

It is often advantageous to collect samples from a boat. In this case it is imperative that life vests be worn at all times. This will not be an obstacle to collecting samples and is probably required for your institution's insurance program.

There must be an experienced captain in charge of the boat. The captain should be someone other than the course instructor, who should be free to advise in sample collection and making measurements such as temperature, conductivity, dissolved oxygen, and salinity. Also, an experienced student is very useful in teaching the use of sampling equipment, especially the ponar grab, which is unwieldy. This assistant can also help get the boat and supplies ready, as well as helping clean the boat and equipment after the collecting trip.

II. Cleanliness, Cleaning Glassware, and Cleaners

Cleanliness is essential in all chemistry courses. Not only is it important in obtaining good results, but it is also important for safety. In environmental studies, scrupulous cleanliness of all equipment that comes in contact with the sample is critically important since the analytes of interest may be present at levels of parts per million, parts per billion, or even lower. It may seem strange to clean equipment scrupulously and then collect a sample of dirt with it. However, if you are looking for trace metals in the soil sample, the collection device must be devoid of even the smallest amount of that metal.

There are exacting protocols for cleaning glassware and other equipment used for various types of tests. When analyzing for trace levels of hydrocarbons, final rinses of glassware with pesticide-grade hexane is recommended; if trace metals are to be studied, a final rinse with dilute, high-purity nitric acid is important.

Generally the cleaning procedure should begin with glassware that is not overly dirty or pitted (to avoid adsorption of impurities). An initial cleaning is done that may simply be rinsing away adsorbed dirt or chemical salts. Soaking with hot or warm dilute phosphate-free cleaner usually suffices to initially clean lab glassware. If necessary, brushes may be used, but care should be taken to minimize damage to glassware surfaces. If more extreme measures are necessary, it may be better to use a cleaner piece of glassware to start with.

After the initial cleaning, specific measures may be used if the glassware is to be used for a specific kind of test. If tests are going to be carried out for chlorinated pesticides, avoidance of all chlorinated solvents is critical. Rinsing glassware with hexane may rid the glass of nonpolar organic pesticides that may interfere in the proposed analysis.

If cleaners or chemicals are used for cleaning, these substances are removed after use by first rinsing thoroughly with tap water, followed by deionized water (DI), and ultimately very high-purity water (18 megohm resistance, or better), if this is deemed necessary. Glassware should drain dry in a dust-free environment. Glassware for certain types of organic analyses may require heating in an oven to remove the last traces of organic compounds.

A great deal of time can be spent just cleaning labware. To avoid this, you should be issued fairly clean labware at the start of the term. To ensure that future students also begin with clean equipment, clean all your labware at the end of the term. Once your equipment is clean, even when it is used for various testing, most of the contaminants on the glass can be simply rinsed away if no precipitates are involved.

General labware cleaning procedure is outlined in Table II-1, which describes the type of labware, what it will be used for, the procedures used to clean it, and storage. The numbers in the table refer to specific, sequential operations, which are listed in Table II-2.

Table II-1 General Labware Cleaning

Labware Description	Analysis	Cleaning Procedures	Storage
Volumetric flasks, pipets, beakers, funnels, graduated cylinders, Erlenmeyer flasks, etc.	Metals	1,2,3,4,7,8,4	Cabinet
	Nutrients and wet chemistry	1,2,3,4,5,7,4	Cabinet
	Hydrocarbons	1,2,3,4,6,9	Cabinet

In the past a mixture of chromic and sulfuric acids was used for cleaning glassware. However, chromium(VI) is a carcinogen, and concentrated sulfuric acid is extremely dangerous. A substitute, Nochromix, uses an inorganic oxidizer not based on chromium, but still uses concentrated sulfuric acid and should not be used unless absolutely necessary, and then only under close supervision.

Table II-2 Key to Cleaning Labware for Critical Environmental Testing

1. Remove labels using soapy water and a blade.
2. Soak briefly in warm laboratory cleaner, several of which are listed in Table II-3.
3. Thoroughly rinse with hot tap water.
4. Thoroughly rinse with DI water.
5. Thoroughly rinse with pesticide-grade acetone.
6. Thoroughly rinse with pesticide-grade hexane.
7. Rinse or soak using 50% nitric acid.
8. Store inverted or capped with a suitable material or stopper.
9. Heat in a drying oven at 150°C or greater.

Recommended General Cleaning Procedure

1. Begin with a dilute detergent solution, using a brush, if necessary, to loosen residues. If the initial cleaning is not satisfactory, soak in hot cleaner for a short time. Laboratory cleaners are listed in Table II-3.

2. Try a stronger cleaner of a different type, and soak in the hot cleaner for an extended time.

3. Rinse several times with tap water, followed by small portions of DI water.

4. For more critical cleaning, follow the instructions given in Tables II-1 and 2.

Table II-3 Laboratory Cleaners and their Properties

Alconox	For organics and residues	Biodegradable	Contains phosphates
Liqui-Nox	For organics, metals, and residue-critical cleaning	Biodegradable	Phosphate-free
Micro	Metals, general cleaning	Biodegradable	No phosphates
Nochromix	Inorganic oxidizer, Leaves no residue	Nonpolluting	Contains concentrated sulfuric acid
Citranox	Metals, scale, fats, and oils—good for trace metal analysis	Biodegradable	Phosphate free; can be used in hard water

III. Waste Minimization and Disposal

Responsibility for Waste Management Guidelines for the regulation of hazardous wastes were established in 1976 by the Federal Resource Conservation and Recovery Act (RCRA). RCRA was enacted by Congress to protect human health and the environment from improper management of hazardous wastes. According to RCRA the generator of a chemical waste is responsible for its management, from its production, to its storage, and ultimate disposal. RCRA regulations are found in 40CFR Parts 260-279.

Laboratory procedures that affect health and safety are governed by OSHA and EPA regulations. OSHA regulations that pertain to laboratory safety and health are found in Title 29, CFR, Part 1910 (29CFR 1910), known as "General Industry Standards." The EPA is responsible for regulations pertaining to waste disposal.

In a university the president is ultimately responsible for the management of hazardous wastes. However, this responsibility is usually delegated to a safety officer or hazardous waste coordinator who manages the hazardous waste program for the university. Mid-level responsibility for enforcing hazardous waste regulations is often given to department or division chairs for their respective areas of authority. However, it is ultimately the responsibility of the laboratory supervisor to insure that all those who use chemicals under their supervision properly handle and dispose of the hazardous wastes they produce.

The functions of the safety officer include:

1. To act as the hygiene officer and/or a source of information on chemical safety

2. To inform the institution about local, state, and federal laws and regulations pertinent to chemical safety

3. To oversee the school's adherence to laws and regulations

If an institution has no hazardous waste program, the initial step is to obtain an EPA identification number. This can be obtained from a regional EPA office or a state hazardous waste management agency (the form required is EPA Form 8700-12). Another initial step is to choose a waste contractor. Licensed contractors can be located through your local environmental agency.

Hazardous Waste A waste is a produced or excess substance that is not going to be reused or recycled. All wastes generated must be characterized to determine if they are hazardous. A waste is classified as hazardous if it is flammable, corrosive, reactive, or toxic in character, as defined by The Code of Federal Regulations (40CFR261). A waste is also hazardous if it is listed in 40 CFR 261.

A. **Characteristic Hazardous Wastes** A waste is hazardous if it demonstrates any of the four characteristics of a hazardous waste mentioned above. Examples of each type follow:

1. Flammable Characteristic Waste
   a) Flammable liquids with a flashpoint less than 60°C (140°F): examples commonly encountered in the environmental chemistry lab include: alcohols (including aqueous alcohol with 24% or more alcohol), benzene, toluene, xylenes, acetone, alkanes, mineral spirits, and many others.
   b) Oxidizers: bromates, chromates, iodates, nitrates, perchlorates, permanganates, peroxides, persulfates (these are especially hazardous when they are in contact with organic matter)

2. Corrosives Aqueous systems having a pH outside the range 2 to 12.5, or other liquids that can corrode steel or burn the skin.
   a) Inorganic acids: frequently used acids are HCl, $HClO_4$, H_2SO_4, HNO_3, and H_3PO_4
   b) Organic acids: formic acid, acetic acid, and the chloroacetic acids
   c) Bases: KOH, NaOH, amines, and silicates

3. Reactive Characteristic Waste These materials can react violently or generate toxic fumes, or the waste contains sulfides or cyanides. Commonly encountered substances of this type are:
   a) Metallic sulfides and cyanides
   b) Metal amides
   c) Tetrahydrofuran
   d) The alkali metals
   e) Perchlorate crystals
   f) Carbonyl compounds

4. Toxic Characteristic Waste This class is a group of heavy metals, pesticides, and organics (Table III-1), which are quantitated by the Toxicity Characteristic Leaching Procedure (TCLP). This procedure determines if a waste has the potential to form a leachate that could cause groundwater contamination. Any detectable amount of these substances must be identified on a hazardous waste label.

Table III-1 The Three Classes of Toxic Substances Based on EPA Testing

Heavy Metals	Organics
Arsenic	Benzene
Barium	Carbon tetrachloride
Cadmium	Chlordane
Chromium	Chlorobenzene
Lead	Chloroform
Mercury	*o* -, *m* -, and *p* – Cresol
Selenium	1,4-Dichlorobenzene
Silver	1,2-Dichloroethane
	1,1-Dichloroethylene
Pesticides	2,4-Dinitrotoluene
2,4-D	Methyl ethyl ketone
Endrin	Nitrobenzene
Heptachlor (and its epoxide)	Pentachlorophenol
Hexachlorobenzene	Pyridine
Hexachlorobutadiene	Tetrachloroethylene
Hexachloroethane	Trichloroethylene
Lindane	2,4,5-Trichlorophenol
Methoxychlor	2,4,6-Trichlorophenol
Toxaphene	Vinyl chloride
2,4,5-TP(Silvex)	

Table III-2 lists toxic characteristic wastes of substances that may be used in this manual. For substances undergoing the TCLP, concentrations that exceed the given values cause the waste to be classed as hazardous.

Table III-2 Toxic Characteristic Wastes

Constituent	EPA Waste Number	Regulatory Level, ppm
Barium	D005	100
Cadmium	D006	1
Chloroform	D022	6
Chromium	D007	5
2,4-D	D016	10
Lead	D006	5
Selenium	D010	1
Silver	D011	5

B. <u>**Listed Hazardous Waste**</u> A substance is hazardous if it is listed as a hazardous waste in the Code of Federal Regulations, 40CFR Part 261. The Code of Federal Regulations is available in libraries or can be obtained from the U. S. Government Bookstore (904-353-0569). The EPA has published four lists with about 800 waste codes for specific and nonspecific hazardous wastes. Most of the listed wastes are hazardous due to one or more of the previous four basic characteristics. Some substances on the lists are acutely hazardous and are indicated by an asterisk. This places the substance into a "P-category," a class devised by the EPA. Special precautions must be used in handling, storing, and disposing of these substances. No compounds of this type are used in the experiments in this manual, and most are not routinely encountered in teaching laboratories.

Small-Quantity Generators and Satellite Accumulation Areas

Universities are usually considered to be small-quantity generators. These are generators of hazardous wastes that produce 100–1000 kg of wastes, or less than 1 kg of acutely hazardous wastes (P-type), in a calendar month. This type of generator must not have more than 6000 kg of hazardous waste in storage. As a small-quantity generator, the university is given an EPA identification number and is required to follow federal and state regulations that cover the management and disposal of hazardous wastes. Regulations require that the university determine if a particular waste is hazardous and, if it is, to manage and dispose of it strictly in accordance with applicable standards.

A satellite accumulation area is a temporary storage and collection area for hazardous wastes close to the point of generation. This area is under the direct control of the supervisor, who is responsible for the waste generated. Each teaching laboratory is considered a satellite accumulation area. Hazardous wastes at a satellite accumulation area can be accumulated as long as necessary, but the amount of all wastes cannot exceed 55 gallons, or 1 quart of an acutely hazardous waste.

Identifying Hazardous Wastes

Although many wastes can be characterized as hazardous or not by using the previous criteria, there are still substances that are difficult to characterize. The following guidelines are useful in characterizing a substance.

1. Refer to the Material Safety Data Sheets (MSDS) that are supplied with the chemical from the supplier.
2. Read the product labels.
3. Refer to the chemical catalog, or confer with the supplier.
4. Confer with your contractor for hazardous waste disposal.

Laboratory supervisors should assume that wastes are hazardous until it is determined otherwise.

Requirements for Handling Waste in Satellite Accumulation Areas

1. Do not accumulate more than 55 gallons of hazardous wastes.

2. Do not use P-listed wastes in teaching labs (authors' recommendation).

3. Accumulate chemical wastes in properly labeled, closed containers. The label should read "Hazardous Waste" and should list all compounds and their percentage. Do not use abbreviations. The start date—when waste was first put into the container—should be given on the label.

4. Hazardous wastes should be stored in secure and safe locations within the generating area. Other chemicals should not be stored in this area.

5. Do not dispose of hazardous chemical wastes in sinks or the trash.

6. Do not use fume hoods to dispose of volatile wastes.

7. Hazardous wastes must be kept in containers that are compatible with the waste. Dry solid wastes should be stored in plastic containers. Also store liquid organic wastes in plastic containers and keep the containers in vented metal cabinets.

8. Collect concentrated acids and bases in separate glass bottles and store apart.

9. Accumulate the following common laboratory wastes in glass containers (but do not combine these substances): aniline, butyric acid, chlorinated solvents, cyclohexane, nitrobenzene, perchloroethylene, toluene, and the xylenes.

10. Solids and liquids are stored in separate containers.

11. Segregate solutions containing heavy metals from other wastes.

12. Do not mix organic solvents and aqueous solutions.

13. Do not mix strong acids or oxidizers with organic compounds.

14. Separate liquid organic wastes as follows:

   a) Halogenated organics: chloroform, carbon tetrachloride, tetrachloroethylene

   b) Polar, nonhalogenated organic compounds: alcohols, carboxylic acids, amines and other hydrogen-bonding compounds

   c) Nonpolar organics: alkanes and aromatics

15. Incompatible chemicals: the following wastes must be kept in separate containers: flammable liquids, acids, bases, oxidizers, halogenated organics, nonhalogenated organics, oils, water reactive substances, mercury and mercury compounds.

16. Physically separate the following categories of hazardous wastes by a barrier or by distance: acids, bases, flammables, oxidizers, and reactives.

17. Mercury

   a) Do not combine metallic mercury with other chemicals.

   b) Store organic and inorganic mercury compounds apart from each other and from other substances.

c) If possible, do not use mercury thermometers (authors' recommendation). If a mercury thermometer is used and breaks, containerize the remains and label the glass and mercury as follows: "Broken thermometer and elemental mercury." Even a small amount of elemental mercury can be a serious health hazard.

18. Have a hazardous waste contractor routinely pick up the wastes to be disposed of in a safe and legal manner.

The Disposal of Chemical Wastes in Sanitary Sewers

Generally, RCRA regulations do not allow any drain disposal except for the following substances:

1. Inorganic and organic acids with a pH between 5 and 10: neutralization is permitted to bring the pH within the proper range.

2. Aqueous buffer solutions that contain no regulated materials: these include common salt solutions (chlorides, bicarbonates, citrates, phosphates, sulfates, and acetates) of sodium, potassium, magnesium, and calcium.

3. Aqueous solutions with less than 24% ethanol, propanol, or isopropanol (24% is based on flammability).

4. No more than unavoidable traces of highly toxic organic chemicals, such as is found on the surfaces of glassware.

There are usually local regulations about what can be disposed of in the drain. The laboratory supervisor should learn the local regulations.

Waste Minimization

Waste minimization is federally mandated for hazardous waste generators. Teaching laboratories should take reasonable and appropriate actions to minimize the amount of hazardous wastes generated in the experiments performed. Techniques that can be used include:

1. Eliminate the waste-generating process. Change the process so that a hazardous waste is not produced. For example, substitute another type of titration for one using silver nitrate, thus eliminating silver chloride as a hazardous waste product.

2. Reuse and recycle materials. If silver nitrate is used in a gravimetric chloride determination, use chemical methods to regenerate silver nitrate from the silver chloride.

3. Substitute a nonhazardous or a less hazardous material for a hazardous material. For example, use a surfactant cleaner in place of chromic acid to clean glassware.

4. Reduce the scale of the procedures or use microscale experiments. In a titration procedure, scale down analyte and use a 10 mL buret instead of a 50 mL buret.

For the experiments in this manual, measures have been taken to prevent hazards normally associated with the handling and disposal of hazardous chemicals. Some measures incorporated in this manual are:

1. Have students work in small groups

2. Stress that students take only what they need from stock bottles

3. Use the same standard solutions for several experiments, if feasible

4. Use the same environmental samples for several experiments

Literature Cited

1. Comprehensive Quality Assurance Plan and Standard Operating Procedures (SOPs), Florida Department of Environmental Regulation (FDER), Tallahassee, FL, 1992

2. EPAs Hazardous Waste Requirements: For Small Quantity Generators of 100 to 1000 Kilograms per Month, EPA/530-SW-86-003, Washington, DC, March, 1986.

3. A Guide on Hazardous Waste Management for Florida's Laboratories, Prepared by Florida Department of Environmental Protection, Hazardous Waste Compliance Assistance Program and Florida Center for Solid and Hazardous Waste Management, Tallahassee, FL, 1997.

4. *Laboratory Waste Management: A Guidebook, ACS Taskforce on Laboratory Waste Management*, American Chemical Society, Washington, DC, 1994.

5. *Less Is Better: Laboratory Chemical Management for Waste Reduction,* 2nd ed., ACS Task Force on Laboratory Waste Management, American Chemical Society, Washington, DC, 1993.

6. Prudent Practices for Handling Hazardous Chemicals in Laboratories, Committee on Hazardous Substances in the Laboratory, Assembly of Mathematical and Physical Sciences, National Research Council, National Academy Press, Washington, DC, 1981.

7. Code of Federal Regulations 40: 260-271, 1998.

8. G. J. Hathaway, N. H. Proctor, J. P. Hughes, and M. L. Fischman, Eds., *Proctor and Hughes' Chemical Hazards of the Workplace, 3rd ed.,* Van Nostrand, New York, 1991.

9. David A. Pipitone, Ed., *Safe Storage of Hazardous Chemicals, 2nd ed.,* Wiley, New York, 1991.

10. R. Scott Stricoff and Douglas B. Walters, *Laboratory Health and Safety Handbook, A Guide for the Preparation of a Chemical Hygiene Plan*, Wiley-Interscience, New York, 1990.

11. Jay A. Young, Editor, *Improving Safety in the Chemical Laboratory, A Practical Guide, 2nd* ed., Wiley, New York, 1991.

IV. The Analytical Balance and Weighing Procedure

Weighings done in the chemistry lab are of two kinds: rough weighing and accurate weighing. Rough weighings are done using a triple beam or top loading balance that can give two to three significant figures. These balances have a large capacity and are used when accuracy is not critical. By definition, an analytical balance can weigh with a precision of at least one part in 10^5 at its maximum capacity. The analytical balance gives at least four significant figures for masses of 100 mg or more. The modern analytical balance is a single pan, direct reading electronic balance that weighs objects rapidly. Electronic balances have a taring mode that causes the display to read zero with a container on the pan.

The electronic balance operates with a constant load, thus giving a constant sensitivity for all masses measured.

General Rules for Using the Modern Electronic Balance

1. Bring the object to be weighed to room temperature. This is frequently done in a desiccator.

2. Brush the pan with a soft brush to clear it of any particulates.

3. Close the doors of the balance and check the zero point. If the balance does not read zero, use the rezero and recheck. If it still does not read zero, see the instructor.

4. Use wipes, finger cots, or crucible tongs to handle containers that are to be weighed to avoid oils from the fingers having an effect on the mass of the container.

5. Place the object to be weighed on the center of the pan, close the balance doors, and wait for a constant reading on the scale. Record the reading in your lab notebook.

6. Remove the object from the balance and close the doors.

7. When weighing solid chemicals, use a weighing boat or a small beaker to hold the substance. Solids can also be weighed by difference using a weighing bottle. The bottle with the substance is weighed, the sample is removed from the bottle, outside the balance, and the bottle is reweighed. The sample mass is the difference between the two masses.

V. The Use and Care of Laboratory Glassware

It is important to stress that the quality of any sampling and analysis activity is directly related to the cleanliness of all glassware and other equipment that the sample may contact. The glassware cleaning procedures described in this manual are simple in order to conserve laboratory time. Specific regulatory agencies and environmental programs often require stringent cleaning criteria when the data generated are to be formally reported.

There are three basic items of precision volumetric glassware: the volumetric flask, the pipet, and the buret. Although you may have used these previously, we shall consider their use and limitations in greater detail.

Accuracy and Precision of Volumetric Glassware

The precision of an item of volumetric glassware is expressed by the uncertainty of a reading. The accuracy of the volumetric glassware is expressed as the maximum allowable error, which is called its tolerance. Table V-1 gives the National Institute of Standards and Technology tolerances for volumetric glassware.

Table V-1 Tolerances of Class A Volumetric Labware

Capacity (mL)	Volumetric Flasks (mL)	Volumetric Pipets (mL)	Burets (mL)
5	—	0.01	0.01
10	—	0.02	0.02
25	0.03	0.03	0.03
50	0.05	0.05	0.05
100	0.08	0.08	0.10
500	0.15	—	—
1000	0.30	—	—

If a 500 mL volumetric flask is used, its volume can be expressed as 500 ±0.15 mL or, in terms of significant figures, a volume of 500.0 mL since the first decimal place is uncertain (see Experiment 1 for a discussion of significant figures). On the other hand, for a 50 mL buret a volume of 25 mL can be written as 25 ±0.05 mL and should be written as 25.0 mL since the first decimal place is uncertain. Although there is uncertainty in the first decimal place of a Class A buret, the buret volume can be estimated to the second decimal place and thus **delivered volumes should be written as 25.00 mL** in the given example. The volume actually lies between 24.95 and 25.05 mL, implying a volume of 25.0 mL.

Glassware with the tolerances given in Table V-1 are said to be Class A. Class B glassware has two times the tolerance of Class A glassware. Class A glassware should be used in all experiments in this manual and any time a high degree of precision and accuracy is required.

A. Use and Care of Volumetric Flasks

Volumetric flasks commonly come in sizes from 10 mL to 2 L, although other sizes are available. They are routinely used to make up standard solutions (solutions of accurately known concentration), laboratory reagents and dissolved environmental samples. They should not be used to store solutions since strong acids, bases, and other corrosives can attack and permanently damage the glassware surfaces.

Provided these flasks are not heavily contaminated, they can simply be rinsed free of the previous solution and rinsed briefly with glassware cleaner, followed by rinses with tap and then DI water. To test flasks and other volumetric glassware for cleanliness, add DI water and let it drain. For a volumetric flask, swirl and invert to drain. If the water does not bead or form streams, the glassware is clean. If not, use mild detergent first, followed by more stringent cleaners; stubborn contamination may require shaking in hot cleaner. Hot cleaner is superior to cold cleaner. The clean, drained flask can be used directly without drying.

To transfer solids from a weighing boat, or a small beaker, to a volumetric flask, pour the powder slowly through a powder funnel in the top of the flask. Rinse the container at least three times with small portions of DI water, and add the rinsings to the flask. Alternatively, first dissolve the sample in the weighing container, and then transfer the solution to the flask. Rinse the container three times with small portions of DI water, adding the rinses to the flask.

After transferring the sample to the flask, add DI water to partially fill the flask. Swirl to dissolve the sample. If the sample is difficult to dissolve, fill to the mark, add a magnetic stir bar, and stir until the sample dissolves. Never heat a volumetric flask because it may break.

Diluting to the mark with water can be done by using a squeeze bottle. Fine addition of water can be achieved by using a dropper to add the last few drops of water.

If a magnetic stir bar is not used to obtain a uniform solution, invert the stoppered flask and move in a circular motion. Continue to invert and mix several times. It is very important that the solution be thoroughly mixed.

Volumetric flasks are normally designed to contain (TC) the stated volume and thus should not be used to transfer a given volume of solution or liquid to another container.

B. Use and Care of Pipets

Pipets are used to accurately transfer known volumes of liquid from one container to another. There are two general types of pipet, volumeric and measuring. A volumetric, or transfer pipet, delivers a single volume, commonly 5, 10, and 25 mL. Measuring pipets, on the other hand, are graduated and can deliver variable quantities of liquid.

A small amount of liquid remains in the pipet tip when the pipet is emptied. If the pipet is TD (stamped on its stem) it delivers the indicated volume and the liquid in the tip must not be blown out. If the pipet is TC (to contain), the residual liquid should be blown out.

To test a pipet for cleanliness, fill it with DI water and allow to drain. If water beads or streaks, the pipet must be cleaned before use. If an automatic pipet cleaner is not available, fill the pipet with warm cleaner and allow to soak for 10 minutes. Allow the pipet to drain, rinse with tap water and then DI water, and again test for cleanliness. If it is not clean at this point, try a stronger cleaner and soak for a longer time.

When ready for use, the pipet must be rinsed with the solution to be pipetted. This avoids dilution of the solution. Draw up a small amount of solution into the pipet, remove the bulb and keep the liquid in the pipet by using a finger on the end of the stem. Tilt the pipet to the horizontal and rotate to rinse. Allow some of the liquid to drain through the stem and the rest to drain through the tip. Rinse two more times.

Fill the pipet by using a pipet bulb to draw liquid about one inch above the line on the pipet stem. Quickly remove the bulb and put a finger over the stem opening to contain the liquid. Release a little of the pressure and allow the pipet to drain slowly to the line. Use a wipe to dry the tip of the pipet. Place the tip of the pipet on the inside wall of the container into which the liquid is to be transferred and allow the liquid to drain. After use, simply rinse the pipet three times with DI water. A squeeze bottle can be used to rinse the insides of the pipet by squeezing water through the stem. If the pipet is allowed to dry while stored in a vertical position, it will be ready for the next use and will not need rinsing, except possibly with the solution to be pipetted.

Pipets are used to take aliquots of solutions—either a solution that is to be analyzed or a standard solution that is to be diluted. An aliquot is an accurately-measured part of a solution. To avoid contamination, do not dip the pipet tip directly into a test or standard solution, but use a small beaker to hold a part of the solution that is to be pipetted. For your own safety, and to avoid contamination of the solution being sampled, use a bulb, never your mouth, to draw liquid into the pipet,

Finally, a note should be added about micropipets, which are used to transfer variable microliter quantities of liquid. They are very useful in preparing standard solutions for several analytical methods, especially for very dilute solutions. However, given the small volumes and the variability of pipettors and tips, their calibration prior to use is critical to maintain acceptable levels of precision and accuracy.

C. Use and Care of Burets

For convenience, burets having Teflon stopcocks should be used as they require no lubrication. Check for cleanliness by filling the buret with DI water and allowing the buret to drain. If water does not bead or streak on the glass, the buret is clean. Otherwise, clean by filling with dilute detergent, allowing the buret to soak for 15 minutes. Drain and rinse thoroughly with DI water and again test for cleanliness. If the buret is still not clean, use a stronger cleaner and soak longer.

Before use, the buret must be rinsed with the solution to be used for the titration. To accomplish this, and to do a titration, you should have a small waste beaker and a beaker for adding titrant to the buret. You may also wish to use a funnel in the top of the buret for adding titrant. Close the stopcock and add about 10 mL of titrant to the buret and tilt the buret to the horizontal and rotate to rinse the sides. Open the stopcock and allow the buret to partly drain into the waste beaker. Then allow the rest of the contents to drain from the top of the buret. Repeat twice.

Fill the buret above the zero mark. Open the stopcock completely to get rid of air bubbles in the tip. Close the stopcock and refill to above the 0.00 mL mark, if necessary. Slowly open the stopcock and allow the liquid to drain until the bottom of the liquid meniscus is at the 0.00 mL mark, then begin titration.

To read the buret, bring the meniscus to eye level by adjusting the buret in the buret clamp. A magnifier helps you read the position of the meniscus. A hand or white paper may aid in reading the level of the meniscus.

When the end point is approached, rinse down the sides of the titration vessel with DI water from a squeeze bottle. This brings splashed analyte back into the bulk of the solution so it can be titrated. Near the end point a fraction of a drop can be added by very slowly opening the stopcock and allowing a portion of a drop to form at the tip of the buret. Close the stopcock and rinse off the partial drop using a DI squeeze bottle. When the end point is reached, record the titration volume.

When the titration is complete, drain the buret into the waste beaker and then, with the stopcock open, squeeze DI water down the inside walls using a squeeze bottle. Repeat at least three times. Allow the buret to drain completely and then invert with the stopcock open and clamp with the buret clamp to dry. The next time the buret is needed it can be used without rinsing with the titrant.

VI. **Blanks and Controls**

Blanks provide matrices that possess negligible, or unmeasurable, amounts of the analyte. Sampling blanks for environmental analyses include trip, field, matrix, and equipment blanks. Other useful blanks include: method, preparation, solvent, reagent, and instrument blanks. We shall consider only a few types of blanks, although several are important in highly reliable studies.

Field blanks are samples of the media that are free of analyte and have a matrix similar to that of the sample. These blanks measure contamination of the sample that occurs in the process of working with the sample (from collection to transport to storage, sample preparation, and analysis). If the samples are water, the field blanks should be high-purity water that is taken to the sampling location and exposed to the atmosphere, allowing any contamination from the air to be measured. If hydrocarbons in surface waters are being studied, the field blank will detect contamination due to an outboard motor.

Equipment blanks (also called rinsate blanks) are obtained by rinsing the sampling equipment with analyte-free solvent. These blanks detect contamination of the sampling equipment. If the analytes are hydrocarbons, the equipment is rinsed with a nonpolar solvent such as hexane, and the hexane rinse is analyzed for the presence of hydrocarbons.

Background (or control) samples are needed for scientifically comparing samples suspected of containing contaminants with those not containing these contaminants. Background samples are collected and analyzed using the same methods used for the study sample. This establishes the presence and/or concentration levels of the analytes studied and the effects of the matrix on their analysis.

Control site samples are taken close to the same time and near the site of the sample to be studied. These samples demonstrate whether the study site is contaminated or is simply average background. A background control is always necessary for valid, scientific comparisons. A sandy sediment control is necessary if the sediment sample under study has a sandy matrix.

Literature Cited

1. Maria Csuros, *Environmental Sampling and Analysis for Technicians*, Lewis Publishers, Boca Raton, FL, 1994.

2. Lawrence H. Keith, Ed., *Environmental Sampling and Analysis, A Practical Guide*, Lewis Publishers, Boca Raton, FL, 1991.

3. Lawrence H. Keith, Ed., *Principles of Environmental Sampling*, American Chemical Society, Washington, DC, 1988.

Experiment 1

Statistical Treatment of Raw Data and Properties of Natural Waters

Objectives—The objectives of this experiment are:

1. To introduce the statistical concepts used in environmental chemistry
2. To introduce the use of the analytical balance
3. To introduce the methods used in sample collection
4. To apply statistical concepts to real data, including the density of natural water
5. To use a collected water sample to determine total suspended solids (TSS) and dissolved solids

Introduction—This experiment is designed to give you experience in the statistical treatment of experimental data. Although you may have had some exposure to statistics, its use in the interpretation and validation of environmental data is often critical for evaluating the significance of apparent results. Thus, a review of some of the important statistical parameters is necessary. The words or phrases used here may sound similar to those from other courses, but there are often subtle differences in meaning and/or use.

When evaluating data there is often an attempt to classify the data as "good" or "bad." These are subjective terms and do not provide any measure of data quality. For scientific and legal purposes, data quality is measured by the adherence to specific requirements of established, proven methods used to collect, prepare, and analyze sample(s) and generate the final results. Further, reported results must be shown to meet the statistical criteria established for all procedures employed to produce them.

For the end user of the data, there is an additional measure of data quality. Do the data meet the data quality objectives established prior to sample collection and analysis? Data quality objectives may include a few or many criteria, such as instrument and/or method detection limits, special sampling, or analytical requirements. Much of the environmental data produced is required by laws that protect the environment. Thus local, state, and federal permits and programs often determine data quality objectives.

Theory

Part A: Concepts and Practice of Statistics

Steps in an Analysis

There are three steps in establishing the results of an analysis. In Step 1 we record, in an appropriate manner, the data as they are obtained. Recording experimental data is usually done by writing the observations, as they are made, in ink in a hardbound laboratory notebook. However, a tape recorder can be used initially and the data eventually transferred to a notebook. At times, photographs should be taken and copies included in the notebook. Various calculations, leading to final results, are also shown.

In Step 2 we decide the best value of the results to report. Usually multiple measurements (replicates) are obtained for a given sample. A soil sample from a landfill may yield the following results for chromium: 20.8, 20.2, and 15.7 ppm. The analyst must decide on a value to report that best characterizes the sample under study. The value reported is often the mean value.

In Step 3 we indicate the precision (scatter) of the results. This indicates the homogeneity of the samples, the appropriateness of the method for the sample, and also the care and skill of the experimenter.

We shall discuss and quantitate the previous concepts and shall give practical methods for dealing with small sets of data and outliers—values that differ greatly from the other values.

Best Value

The value reported is frequently the arithmetic mean, although the geometric mean is useful if there is an outlier. Another important reported value is the median.

Arithmetic Mean or Average—This best value is the sum of the individual measurements divided by the number of measurements, mathematically given by Equation 1-1:

$$\overline{X} = (X_1 + X_2 + \ldots + X_N) / N = \sum X_i / N \qquad \textbf{(1-1)}$$

where $\overline{X}$ is the mean, the X_i are the individual results, and N is the total number of results.

Geometric Mean—This is sometimes the preferred value to report as it does not count equally a value that is statistically significant, but which is far removed from the remaining values (an outlier). For example, if five samples of sediment are collected, it is possible that four of five measurements of the manganese content will be comparable and one value will be considerably different. In this case, a geometric mean is preferable to the arithmetic mean. Mathematically, the geometric mean is given by Equation 1-2:

$$X_{geo} = (X_1 \cdot X_2 \cdot \ldots X_N)^{1/N} = [\Pi(X_i)]^{1/N} \qquad \textbf{(1-2)}$$

Median—This measure is simply the "middle" value. When all results are listed in order of increasing value, it is the middle result if the number of values is odd. If the number of results is even, it is the average of the two middle values. The median is not used as often as the arithmetic mean.

Accuracy

Accuracy—how close a result or best value is to the true value. The true value is the exact answer or result of an analysis. This value is often unknown. The use of carefully prepared and analyzed standards will produce a value that is often used as a true value (synonyms: accepted, actual, authentic, right, and correct).

Result—The final product of an analysis, (synonyms: answer, etc.).

Measurements of Accuracy

Absolute Error—The absolute value of the difference between an individual result and the true value, $|X_i - \mu)|$, where μ is the true value.

Relative Error—The absolute error divided by the true value. Often expressed as percent or parts per thousand (ppt) when multiplied by 100 or 1000, respectively. The relative error is equal to $|(\overline{X} - \mu)| / \mu$.

Precision

Precision—the closeness of the results in a set of replicate analyses to each other.

Measurements of Precision

Range—The "spread" of the results. The largest value minus the smallest value is the range (w) of a set of measurements.

Relative Range—The range divided by the mean value for the data set. This value is often expressed as a percentage.

Deviation—The absolute value of the numerical difference between a given result and the mean (analogous to the absolute error), $d_i = |X_i - \overline{X}|$.

Average Deviation—The average of the individual deviations, $\Sigma d_i / N$, where N is the total number of replicates.

Uncertainty

Measurements made using an instrument are subject to some uncertainty due to estimating the position between graduations. For an analytical balance the uncertainty in a measured mass is at least ±0.0001 g. Thus, a 5.5512 g mass can be between 5.5511 and 5.5513 g.

A reasonable uncertainty is ±½ the distance between the smallest graduations. A buret usually has graduation marks every 0.1 mL, and the liquid level between marks can be estimated to no better than the nearest 0.01 mL for an experienced analyst.

Uncertainty is the "precision" of a single measurement. Even a digital readout has an uncertainty. When you look at the illuminated numbers of a digital readout, you usually see small fluctuations in the last digit. The uncertainty in a measurement is taken to be ±1 in the final digit.

Uncertainty expressed in the units that are measured is called the absolute uncertainty. The relative uncertainty is the absolute uncertainty divided by the number measured multiplied by 100 or 1000 to give the relative uncertainty in percentage or ppt.

Error Analysis – Types of Errors

The interpretation of results that are not exact requires an analysis of errors. A report of experimental results must include a discussion of errors observed or inferred from the data.

Experimental measurements are affected by two principal types of error. Random errors, also called indeterminate errors, result in deviations that may be either positive or negative. Random errors cause data to be spread somewhat symmetrically about the mean value if there are no other errors present. It is often difficult to ascribe exact causes to random errors; however, much research has been done to minimize random errors in analytical instruments.

If errors are truly random, it is possible to approximate a true value by using the average measurement for a sufficiently large number of samples (or analyses). To determine whether "enough" analyses have been performed, a few more are carried out and the average calculated after each new measurement. If the average does not change (significantly), the average is acceptable.

The second type of error commonly found is systematic, or determinant, error. This type of error causes the mean of a data set to differ from the true value of the sample. Generally, a systematic error causes the results of replicate analyses to be consistently high or consistently low.

A third type of error frequently encountered, especially in dealing with environmental samples, is gross error. A gross error results in an outlier that is very different in value than the remainder of the results. This type of error may be due to an inhomogeneity in the sample (poor sampling), the presence of a contaminant, or making a mistake in reading a buret or balance.

Random Errors and the Distribution of Experimental Results

The cause of a gross error can be determined and eliminated, although this may not be easy in practice. Systematic errors also can be located and eliminated. If a high buret reading is constantly made or if an indicator change is not intense enough to be seen, results will be consistently high for the analysis. The errors can be eliminated, however. Another systematic error is due to a slow titration reaction and can be eliminated by heating the titration mixture. On the other hand, random errors cannot be eliminated and result in a spread, or distribution of results symmetrically distributed about the mean value.

To illustrate the effect of random errors on results, consider the following examples. First suppose that an analysis is carried out carefully so that only random errors occur. To start with the simplest possible situation, imagine that there are just two indeterminate errors in the experiment. An example of this type is reading a buret two times to obtain the volume of a titrant. If the magnitude of the random error is constant, and equal to 0.05 mL, there are three possible results: (1) both errors are positive, giving a total random error of +0.10 mL; (2) both errors are negative, giving a total random error of –0.10 mL; or (3) one error is positive and one error is negative, giving a random error of 0.00 mL.

There is just one way the error can be +0.10 mL and that is for both errors to be positive. The same is true for the error of –0.10 mL. The error of 0.00 mL is two times as probable since there are two ways it can occur: (1) the first measurement being high by 0.05 mL, the second measurement being low by 0.05 mL and (2) the first measurement being low by 0.05 mL and the second measurement being high by 0.05 mL. Thus, if only random errors are present, errors in measurements tend to even out.

Let's now consider a more complex situation, one where there are three equal random errors. We shall consider this to result from reading a buret two times and estimating the equivalence point in a titration. The possible errors, and the ways in which these can be achieved, are shown below. An arrow pointing upward represents an error of +0.05 mL and an arrow pointing downward represents an error of –0.05 mL.

↓↓↓	↓↓↑	↑↑↓	↑↑↑
– 0.15 mL	↓↑↓	↑↓↑	+ 0.15 mL
	↑↓↓	↓↑↑	
	– 0.05 mL	+ 0.05 mL	

Another way to express these results is to plot the number of ways each error can occur versus the value of the error. This is shown in Figure 1-1, where the curve is the expected distribution for a larger number of random errors.

Figure 1-1 Frequency of Errors Versus Value of the Error for Three Equal Errors

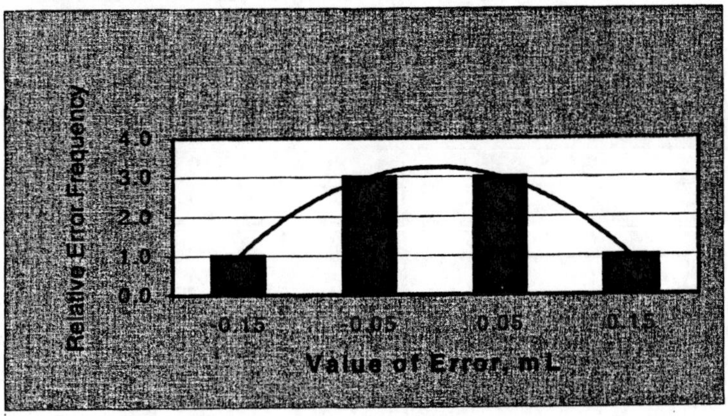

Before we examine the ideal curve that represents the distribution of results when there are only random errors, let's take another type of example that shows the statistical nature of random errors. If 10 playing cards are drawn from a well-shuffled deck of cards, the probability of drawing five red suits is 0.5 or 50%

(26 red and 26 black). However, this will be true only if a sufficiently large number of trials is carried out (in statistics, this is called a population). We shall examine the results of drawing cards 10 times from a shuffled deck, reshuffling and again drawing 10 cards. This is done a total of 10 times so that 100 cards are drawn in all for one trial. Then the process is continued for a total of 10 trials and the results are shown in Table 1-1.

Table 1-1 The Number of Occurrences of Red Suits Drawn from a Deck of Cards

Draw	Trial									
	I	II	III	IV	V	VI	VII	VIII	IX	X
1	4	6	4	4	5	7	5	4	5	3
2	8	6	6	6	6	5	4	4	5	3
3	4	4	2	4	8	9	5	6	3	4
4	7	6	4	4	3	6	6	6	6	5
5	7	5	7	4	3	6	5	5	5	4
6	6	6	6	5	6	5	5	4	3	6
7	2	6	5	6	5	3	4	7	5	7
8	7	8	4	4	4	6	2	5	6	3
9	4	6	2	4	5	5	3	6	7	4
10	7	3	5	6	4	5	3	7	3	7

The frequency with which different numbers of red suits occur is listed in Table 1-2. The listings in this table indicate the number of times 0 red suits show up, the number of times just 1 red suit appears, and so on.

Table 1-2 Frequency of Occurrence of Different Numbers of Red Suits for 10 Draws

Number of Red Suits	Observed Frequency
0	0
1	0
2	4
3	12
4	23
5	22
6	24
7	11
8	3
9	1
10	0

A scatter plot made from the data of Table 1-2 is shown in Figure 1-2. As in Figure 1-1 an ideal curve is drawn that would fit the data if there were an infinite number of a continuum of results. As long as a system is statistical (random and large enough), the curve will visually express the distribution of errors about the mean value, μ, which is also the actual value in this instance.

Figure 1-2 Frequency of Occurrence of Red Suits Versus Number of Red Suits per 10 Draws

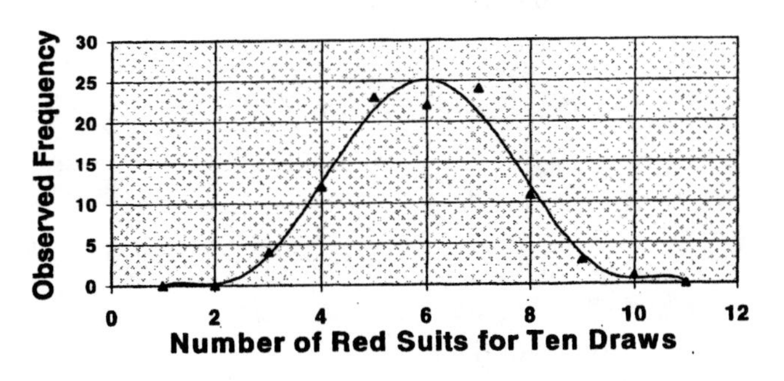

To understand the importance of this ideal curve, which is called Gaussian, or a normal distribution (the curve is called a bell curve and is how grades are ideally supposed to be distributed), we first examine its mathematical form. Figure 1-3 shows Gaussian curves in which the relative frequency of occurrence of various results are plotted (along the y-axis) as a function of the actual results (plotted along the x-axis). The curves are described by an equation having two parameters, the mean of the population, μ, and the standard deviation, σ.

$$y = \{\exp[-(x - \mu)^2 / 2\sigma^2]\} / \sigma\sqrt{2\pi} \qquad \textbf{(1-3)}$$

In Figure 1-3 we show the results of analyzing a water sample known to be 40.0 ppm in calcium ion. The standard deviation for one set of analyses (Sample A) is 0.10 and for Sample B the standard deviation is twice as great. Observe the symmetry of the curve that results from the squared term in x, and also note the exponential relationship of y to the deviation from the mean. This indicates that the frequency of results with large deviations from the mean is small and that most results have small deviations.

If the number of results is infinite (in reality, more than 20–30), the population mean μ is equal to the true value for the measured quantity. When the number of results, N, is small, the replicate observations are called a sample, and the mean value $\overline{X}$ is defined by Eq. 1-1. In this instance $\overline{X}$ differs from μ and their difference decreases as N approaches 20–30.

The standard deviation for a population measures the precision of a population and is defined by

$$\sigma = [\textstyle\sum(x_i - \mu)^2 / N]^{1/2} \qquad \textbf{(1-4)}$$

The standard deviation determines the breadth of the curves shown in Figure 1-3. It is thus an indicator of the scatter of the data. The precision of the data leading to curve A of Figure 1-3 is twice as good as the precision of the data leading to curve B.

Irrespective of the width of the curves in Figure 1-3, it can be shown that 68.3% of its area lies within $\pm 1\sigma$ of the population mean. Also, 95.5% of the data are within $\pm 2\sigma$ and 99.7% are within $\pm 3\sigma$ of the population mean. Thus we can say that a result lying outside 2σ has only a 4.5% probability of being statistically significant.

If only a sample (small data set) is taken, Eq. 1-4 is no longer valid and if used, will give an estimated standard deviation that is too small. To obtain a better estimate of the standard deviation, Eq. 1-4 is modified to:

$$s = [\sum(X_i - \overline{X})^2 / (N-1)]^{1/2} \qquad \textbf{(1-5)}$$

where μ from Eq. 1-4 has been replaced by $\overline{X}$ and N by $N-1$ (called the number of degrees of freedom).

Figure 1-3 Relative Frequency of Results Versus Value of Results for Two Standard Deviations

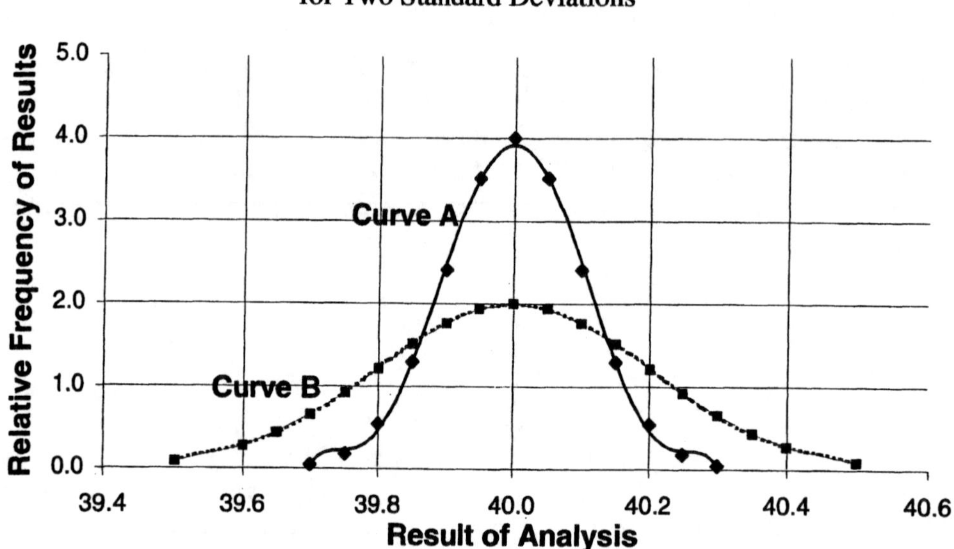

Many scientific calculators have a function key for calculating the standard deviation. You should use your calculator instructions to learn how to calculate a standard deviation using your calculator.

Generally, the experiments carried out in this manual call for determinations in triplicate. At least three results should be obtained to justify the use of Eq. 1-5 in estimating a standard deviation.

Of more use as a measure of precision is the <u>relative standard deviation</u>, expressed in ppt:

$$\textbf{Relative Standard Deviation} = (s / \overline{X})\ 1000\ \textbf{ppt} \qquad \textbf{(1-6)}$$

This property is the standard deviation divided by the mean. When expressed as a percentage it is called the <u>coefficient of variation</u>, CV, given by:

$$\text{CV} = (s / \overline{X})100\% \qquad \textbf{(1-7)}$$

Both of these quantities allow a comparison of one set of data with another (a classmate's, for example).

The Difference Between Accuracy and Precision

To illustrate the difference between accuracy and precision, consider the results of four sets of experiments shown in Figure 1-4. A sample is made up to contain exactly 75.0 ppm of calcium ion, Ca^{2+}. The intersection of the horizontal and vertical lines in the figure represents the true value.

Analyst A not only obtained accurate results, the mean of which was equal to the actual value, but this perfectionist was also precise. There is little scatter to these data, and the deviations are likely a result of small, random errors.

Analyst B obtained precise results but his or her mean differed significantly from the true value. The good precision indicates small random errors. All values appear to be "wrong" in the same direction, indicating a systematic bias, or error. When results are consistently high, a possible systematic error is the presence of an interference. If this analyst used a deionized water bottle contaminated with tap water, which is high in calcium, this would explain the consistently high results.

Analyst C is lucky. Although his or her precision is terrible, the average value of the results is not far from the true value. A large systematic error is indicated because only the average, not the results themselves, is close to the "true" value. Any interpretation of this data set should address both its accuracy and lack of precision.

Analyst D obtained data that exhibit poor accuracy, poor precision, and an apparent systematic error (all results are skewed to the right side of the plot). Results like these indicate several possibilities: the technique of the analyst is inappropriate to the sample, the sample is inappropriate to the method, the scale of the analysis is too precise for the quality of the results that can be obtained, or there is an interference present.

Good Accuracy and Precision

It is not easy to classify data as very accurate or precise. However, it is possible to use empirical analytical results to obtain an idea of the quality of a set of results. If the very accurate analytical method, gravimetry, is analyzed for the best possible accuracy and precision, the best to be expected, if only random errors are present, is 2–3 ppt relative accuracy and relative standard deviation. On the other hand, for volumetric methods of analysis 2–3% is more usual.

Your results must indicate in some manner or another the reproducibility of the results, the uncertainties in the measurements, and the precision, as indicated by some statistical measure, such as the relative standard deviation. In place of this, or in addition to this, we can also use the significant figure concept to indicate the reliability of a final reported result.

Significant Figures

The significant figures in a number resulting from a measurement, or a result that has been calculated from a measurement, are the digits that indicate its precision. It is the number of digits known with certainty plus the first uncertain digit. The number of significant figures in a mean value indicates the level of precision of the analytical procedure required to obtain that value. (Most general chemistry texts review the use of significant figures in an appendix. See References 1 and 4 in the Literature section.)

For experiments done in this manual, it is expected that there be an indication of possible random and systematic errors. Also, a relative standard deviation should be calculated and discussed and the proper number of significant figures reported in the final calculated result.

Figure 1-4 Two-Dimensional Representation of Values for Four Sets of Repetitive Measurements Indicated by As, Bs, Cs, and Ds

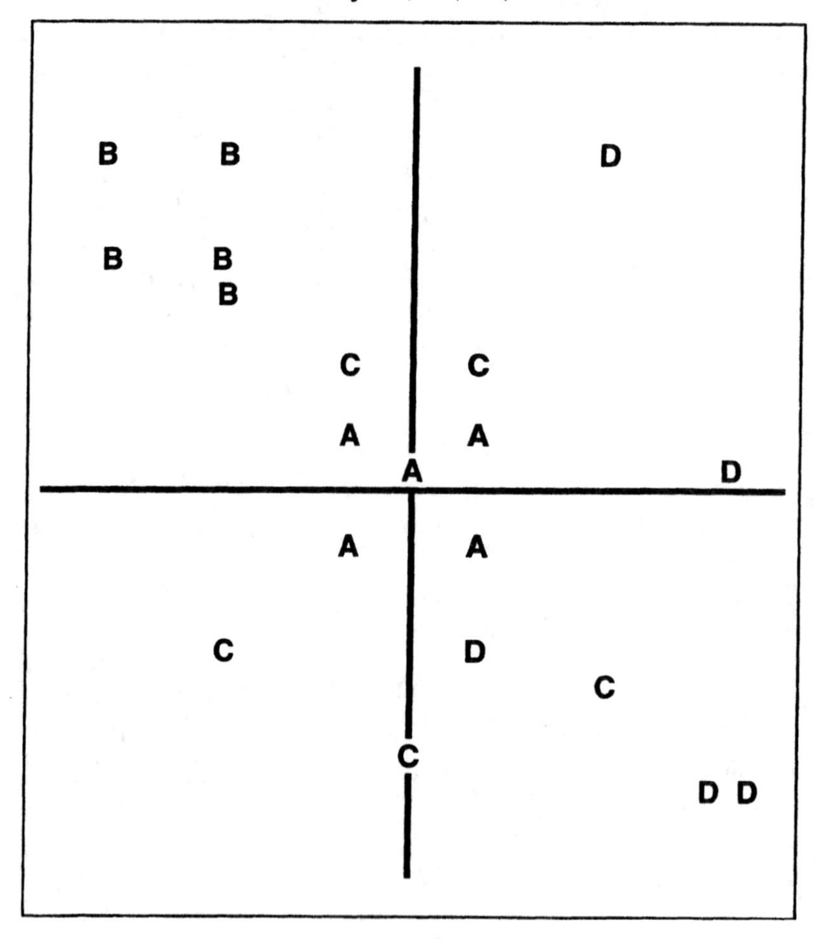

Calibration and Calibration Curves

Before completing this section we shall consider one last but very important application of error analysis, the method of "least-squares." This method will be used in many experiments in this manual and will be encountered in many other areas of science.

It is seldom possible to use an instrument to obtain a usable result directly. For example, if a spectrophotometric method is used to measure the concentration of a solution, the instrument gives a readout of absorbance, an indicator of the amount of light absorbed by the analyte in solution. To obtain the concentration it is necessary to obtain the relationship between absorbance and concentration; this is accomplished through calibration.

The simplest case is a direct proportionality between the measured property and concentration. In spectrophotometry the absorbance and concentration are directly related, a relationship called Beer's law, expressed mathematically by

$$A = k \times C \quad (1\text{-}8)$$

where k is a constant, A is the absorbance, and C is the solution's concentration. If we make up a solution of known concentration and measure its absorbance, we can calculate the constant. Then a solution of unknown concentration is examined to determine its absorbance, and its concentration can be calculated. This is called the standard comparison method.

The method just described is called calibration and is a fundamental and general method used throughout the sciences. In this manual, in addition to spectrophotometric measurements that require calibration, ion selective electrodes and fluorimetry determinations also require calibration. Although a standard comparison can suffice in these measurements, it is more reliable to prepare a calibration curve, which is a plot of the physical property measured (along the y-axis) versus the concentration (along the x-axis).

Often a straight line is drawn manually through the points determined, and the estimated straight line is used to approximate a concentration from the measured property of the solution (absorbance, potential, etc.). Most often there is scatter in the data and a perfect straight line is not obtained. This is a result of random errors, under the best of conditions, but may also include systematic errors, which often are not difficult to recognize because of their large magnitude. There is a better way to determine concentrations from calibration data, and this is the method of least squares.

In a least-squares method the chosen line minimizes the sum of the squared deviations of the individual results. Although this process can be carried out manually, it is tedious and it is easy to make errors. Most calculators have a statistical mode that allows a quick determination of the best straight line for the experimental data. Most computer software for science applications has a least-squares program that also generates the straight-line data. The results of such an analysis are the slope of the line and the y-intercept. The intercept is important since it indicates whether there is a significant blank contribution to the observed property.

A third result of the least-squares analysis is the correlation coefficient, r. This indicates how well the data points fit the straight line—in other words it indicates the scatter of the calibration data, or the presence of determinate errors. A perfect fit has a value of r equal to 1.000 (or $r = -1.000$ for a negative slope). (Some software report a value for r^2 instead of r.)

If computer software is used, a least-squares line can be generated and used directly to estimate the concentration of the solution under study. This still requires some guesswork and a better method is to generate the equation of a straight line, $y = mx + b$, where m is the slope and b is the y -intercept, determined from the least-squares analysis. Then the experimental value of y is inserted into the equation, which is then solved for the concentration, x.

Many of the experiments in this manual require a linear-least squares analysis, and the results of the analysis should be submitted with the laboratory report.

Safety Issues—Safety glasses must be worn whenever you are in a chemistry laboratory.

Procedure

Part A: Using the Statistical Treatment of Data

There are two objectives to this part of the experiment: first, to learn the use of an analytical balance, and second, to obtain a data-set that will allow the calculation of the measures of accuracy and precision discussed previously.

In your lab notebook, outline weighing procedure so that you can refer to it in the future. Have your lab assistant or instructor check you out on the balance before beginning this part of the experiment. Always use the same balance to carry out all weighings in an experiment.

1. Verify that the balance is level. If necessary, use a brush to clear any material from the balance pan. Zero the balance.

2. Balance doors should be closed when weighing an object, and the object must be at ambient temperature, and dry. Without touching directly with your hands, weigh a weighing bottle with its top to the nearest ±0.1 mg. Weigh the bottom and top separately, and add the masses to see if the sum is equal to the mass of the two together. If not, consult your instructor, or try again.

3. Obtain from the instructor about 30 small objects of the same type and material. Many possible objects will suffice, including laboratory articles such as weighing boats, weighing paper, filter paper, etc. Other objects with variable masses include aspirin (or other tablets), rubber washers, dried peas, rubber bands, or paper clips. Zero the balance again and then determine the mass of each of the objects supplied.

Waste Minimization and Disposal—No hazardous wastes are generated in this experiment.

Data Analysis

1. Make a table containing the masses of the weighing bottle bottoms, tops, and both together. Calculate the difference between the sum of the top and bottom masses and the experimentally determined mass of the two together.

2. For the masses of objects weighed, prepare a table listing the masses in order of increasing mass. From the masses, calculate the following: (a) mean mass, (b) median mass, (c) the range of the masses, (d) the relative standard deviation in ppt, and (e) the coefficient of variation, CV.

Supplemental Activity—The student is encouraged to use graphical analysis on a computer to analyze the results. In addition, you are encouraged to prepare your lab report using a word processor.

The 30 objects weighed can be organized in a manner to determine if they satisfy a Gaussian distribution. This is greatly facilitated using computer software.

Theory

Part B: Using the Statistical Treatment of Data—Rejecting Bad Data

The Q-Test. In Part A of this experiment, a simple procedure was carried out to obtain a reasonable number of data points, and the interpretation of raw data was discussed. We have yet to learn how to deal with a "bad" data point. The evaluation of an “outlier” must have an unbiased and statistically based criterion. A frequently used method is the Q-test (Q for quotient). (For a more thorough discussion of this topic, refer to the Literature section at the end of this experiment.)

To carry out the test, the results are first arranged in increasing size. The difference between the value in question and its neighbor is divided by the range to obtain a value of Q, as shown in Eq. 1-9:

$$Q_{exp} = (X_? - X_n) / \text{range} \tag{1-9}$$

where $X_?$ is the questionable value and X_n is the neighboring value. The Q-test is often used at the 90% confidence level. Statistical tables are used to compare Q_{exp} to a value at the acceptable level of confidence (the probability that rejecting the value in question is valid), based on the number of observations in the entire data set. Table 1-3 lists critical values of Q (Q_{crit}) that can be used to accept ($Q_{exp} < Q_{crit}$) or reject ($Q_{exp} > Q_{crit}$) a particular value.

To obtain a "real-world" data set and to gain experience with the Q-test, measurements of the density of a natural water will be made.

The interpretation of data is important for an understanding of the results. Essential to the reporting of the results will be the knowledge of when a single datum should be rejected or when an entire data set may be in error. To establish a basis on which to make some of these decisions, the following section describes some of the properties of natural waters.

Table 1-3 Critical Values of Q for the Q-Test

Number of Observations	Q_{crit} 90% Confidence	Q_{crit} 96% Confidence	Q_{crit} 98% Confidence
3	0.94	0.98	0.99
4	0.76	0.85	0.93
5	0.64	0.73	0.82
6	0.56	0.64	0.74
7	0.51	0.59	0.68
8	0.47	0.54	0.63
9	0.44	0.51	0.60
10+	0.41	0.48	0.57

Source: R. B. Dean and W. J. Dixon, *Anal. Chem.*, 23:636-638 (1951).

Suspended Solids in Natural Waters and the Use of Density to Estimate Dissolved Solids.

The goals of this part of the experiment are to gain experience with the correct procedure for using volumetric apparatus and to learn simple environmental sampling procedures. The objectives will be achieved by determining the density of a sample of natural water.

This experiment, and several others, involve measurements of natural waters that will be collected by the student. Because the results of the analyses and measurements are truly unknown, it is the student's responsibility to locate literature sources (refer to the literature section at the end of each experiment) to help discuss the significance of the results. Also, most experiments have an extensive introduction that will help you discuss your results. In the discussion you should look for the meaning of the results, not just the numbers derived from the experiment.

Natural Waters The total amount of water above, below, and on the surface of the earth is about 1.3×10^{21} kg, about 0.02% of the mass of the earth. Most of the earth's water is either contaminated by dissolved salts (the oceans) or relatively inaccessible (ice and snow, underground sources). Lakes and rivers are the primary source of fresh water, but comprise less than 0.01% of the total water supply. However, even this small fraction represents about 4×10^7 kg per person.

The concentration of dissolved salts in river waters is much lower than in seawater. Seawater contains about 3.5%, by mass, of dissolved solids, and has a density of 1.0245 g/mL at 20°C.

The testing of natural waters often includes the determination of various ions. In addition to chemical measurements, many physical properties of natural waters are routinely measured. These include turbidity, color, density, conductivity, suspended and dissolved solids. In this experiment, the density, suspended and dissolved solids will be determined for a sample of water from a nearby body of water.

Suspended Solids The total suspended solids (TSS) test is one of the most common determinations made in wastewater treatment plants. The test is not intended to measure the concentrations of specific chemical substances, but rather give an empirical estimate of water quality by measuring the amount of suspended foreign materials present. It is determined from the weight gain of a filter after drawing a known volume of water through the filter.

Dissolved Solids Material that cannot be removed by a filter of a particular porosity (typically 0.45 μm) is said to be "dissolved." Many, although not all of these species, are inorganic salts or weak organic acids, which ionize in water, resulting in an increase in electrical conductivity. Thus, the dissolved solids can be estimated by using a conductivity meter.

Density The density of a pure substance is often useful in its identification. Also, the density of mixtures can sometimes be used to estimate the amount of one of the substances present. For example, the density of beer that is being brewed is used to estimate its alcohol content. However, density is of little use in determining the nature of substances in solution.

The greater the density is above the value for pure water, the greater the amount of dissolved solids. The difference between the density of the solution and the pure solvent is the mass of dissolved solids per unit volume:

$$\textbf{TDS (mg/L)} = \textbf{(density of solution – density of pure water) (1000 mg/g) (1000 mL/L)}$$

To obtain a "real-world" data set and to gain experience with the Q-test, measurements of the density of a natural water will be made.

The interpretation of data is important for an understanding of the results. Essential to the reporting of the results will be the knowledge of when a single datum should be rejected or when an entire data set may be in error. To establish a basis on which to make some of these decisions, the following section describes some of the properties of natural waters.

Sampling Samples collected for analysis should be obtained in such a way as to provide the most representative sample possible. In general, samples should be taken near the center of the body of water and entirely below the surface.

It is difficult to obtain a truly representative sample when collecting surface water samples. More meaningful results are obtained by carrying out a series of tests with samples taken from several locations and depths and at different times. The results can then be used to establish patterns applicable to that particular body of water.

Although sample collection will be discussed in Experiment 2, we shall establish some general principles that apply for all types of samples and, in particular, for the water samples collected in this experiment.

Generally, as little time as possible should elapse between collecting the sample and carrying out the analysis. Depending on the nature of the test, special precautions in handling the sample may be necessary to prevent natural interferences, such as bacterial growth or the loss of dissolved gases. Table 2-1 gives detailed information for preserving samples.

When studying a particular aquatic ecosystem, an environmental scientist learns as much about the system as possible. This knowledge helps to explain results and aids in locating areas of the system that should be studied.

Safety Issues

1. Safety glasses must be worn at all times in a chemistry laboratory.

2. When collecting samples, an instructor must be present.

Figure 1-5 A Simple Water Collector

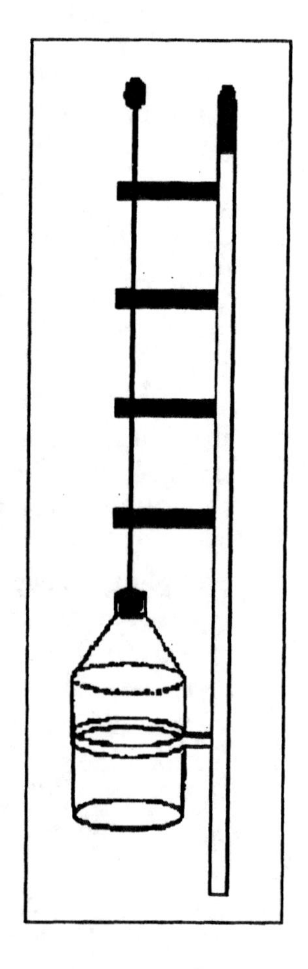

Procedure

Part B: Properties of Natural Waters and the Q-Test

Sampling

Clean a 1 L plastic bottle by washing with 10% hydrochloric acid, followed by thoroughly rinsing with DI water. Collect a sample of water from about 1 m below the surface of a river or pond, from a boat dock, or from another convenient location. One device for collecting water samples is shown in Figure 1-5. In your laboratory notebook record all conditions under which the sample was obtained (air and water temperatures, weather conditions, tide, etc.).

It is good practice to obtain a second sample on a collection trip to serve as a reference. This sample should be collected, handled, and stored the same way as the sample of interest; however, it should be collected in an area that is as pristine (not impacted) as possible. This may be difficult to find in an urban area.

Determination of Suspended Solids

1. For the purpose of this experiment, suction filtration is acceptable. Use a circle of preweighed (to a constant weight) 55 mm Whatman glass microfiber filter paper in a Büchner funnel on a 1 L filter flask and collect 500 mL of filtrate. The filtrate will be used in this and in subsequent experiments and should be stored in a labeled plastic bottle, preferably at about 4°C. In addition to your name, the label should have the collection date and location. In this and the subsequent step, the filter paper should be handled with forceps.

2. Dry the filter on a watchglass in a drying oven at 110°C for one hour. Cool the paper for 1 minute, weigh and heat for another half hour. Reweigh the filter. Repeat the cooling, weighing, and drying until the paper achieves a constant weight (within 0.0002 g).

Your instructor may have you use a desiccator when cooling and storing the filter circle.

Determination of Density

1. Rinse a clean 50 mL buret three times with 10 mL portions of the filtered water sample. Then fill the buret with the water and record the initial volume. (Warm the water to room temperature, if necessary.) Make sure that all air bubbles are absent from the tip of the buret.

2. Weigh four clean and dry weighing bottles, with tops, to 0.1 mg.

3. Add approximately 11 mL of the sample to one of the weighing bottles, 15 mL to the second, and 19 mL to the third. Record the exact volume added to each bottle to the nearest 0.01 mL. Stopper the bottles and weigh to ±0.1 mg.

4. Repeat the last step, measuring out about 9 mL of water, but suppose that a mistake is made by recording 10 mL instead. (Write the true volume in parentheses after the supposed incorrect volume.) This will be used to illustrate the magnitude of the error needed to throw out a questionable sample using the Q-test.

5. Weigh a 25 mL volumetric flask (Class A) and pipet 25 mL of the filtered water sample into the flask. Stopper the flask and reweigh. Note whether the water accurately fills the flask to the mark on the neck of the flask.

6. Weigh a clean and dry 10 mL volumetric flask (Class A) with stopper. Fill to the mark with filtered sample water, stopper, and reweigh.

7. Record room temperature.

Waste Minimization and Disposal—Since the samples collected are not necessarily hazardous wastes, the filtered, suspended solids can be disposed of in a trash receptacle and the water can be disposed of in a drain, if your instructor so directs. However, water samples will also be used in several subsequent experiments.

Data Analysis

1. Report all observations and conditions for the sampling. Include water and air temperatures and weather conditions.

2. Calculate and report the value of the suspended solids, in ppm. Compare with the values given in Table 1-4, or previous values obtained in your area.

3. Prepare a table showing the masses and volumes of each water sample and the calculated density for each sample according to the equation:

$$\text{Density (g/mL)} = \text{mass(g)/volume (mL)} \quad (1\text{-}10)$$

Table 1-4 Previous Results for the Density and Other Water Properties for the St. Johns River, Jacksonville, Florida, Over a 10-Year Period

Parameter	Value
Mean density at 25°C	1.0037 g/mL
Median density at 25°C	1.0035 g/mL
Range of densities	0.0110 g/mL
Relative average deviation	2.8 ppt
Relative standard deviation	3.0 ppt
Mean of suspended solids	8926 ppm*
Mean of dissolved solids	8330 ppm*

*These vary widely with rainfall, the seasons, tides, etc.

4. Use the Q-test at the 90% confidence level to see which, if any, of the density results can be rejected. The most suspect value is the one where the volume was incorrectly recorded. If this value cannot be rejected, calculate the error needed to be able to reject a sample at the 90% level. If any value(s) can be rejected, recalculate all of the quantities of Step 3 after rejecting the suspect value.

5. In another table, present the results of calculating the following: the arithmetic mean, the median, the relative average deviation (in ppt), the relative standard deviation (in ppt), and the relative range for the density measurements. Carry out the calculations for both the case where the volume of one sample was deliberately incorrectly recorded and for the actual volumes, and present all the results in one table.

6. Use the exact value of the density of pure water at the temperature of your density measurements and your mean density for river water to calculate the total dissolved solids (TDS) in ppm. The densities of pure water at various temperatures are given in Table 1-5. The difference between your mean density and the density of pure water at the same temperature is the TDS per mL. Calculate the ppm (mg/L) of TDS. Compare your value with that for seawater and any other available values.

Table 1-5 The Density of Pure Water at Various Temperatures

Temperature	Density (g/mL)	Temperature(°C)	Density(g/mL)
15.0	0.99810	23.0	0.99651
16.0	0.99790	24.0	0.99631
17.0	0.99780	25.0	0.99602
18.0	0.99761	26.0	0.99572
19.0	0.99741	27.0	0.99552
20.0	0.99721	28.0	0.99522
21.0	0.99701	29.0	0.99493
22.0	0.99671	30.0	0.99463

<u>Source</u>: David R. Lide, Editor-in-Chief, *CRC Handbook of Chemistry and Physics*, 59th ed., CRC Press, Inc., Boca Raton, FL, 1996.

Supplemental Activity—In addition to your density measurements for water from a stream, a concurrent study of rainwater or seawater would give contrasting values of the density that could easily be discussed.

Questions and Further Thoughts—If the uncertainty in reading a buret is ±0.02 mL, what is the relative uncertainty (precision) in a measured volume of 25.00 mL in ppm?

Notes

1. The authors have found peanuts to be interesting to use in this experiment as they give a wide variety of masses, and also provide some energy for the trip to collect the water samples.

2. Pennies also provide a good sample type, and if a large range of dates is available, the time when the composition of the penny was changed can be determined.

Literature Cited

1. For more on significant figures, refer to *Chemistry, the Central Science,* by T. L. Brown, H. E. LeMay, Jr., and B. E. Bursten, Prentice Hall (Upper Saddle River, NJ), 1997.

2. J. S. Fritz and G. H. Schenk, *Quantitative Analytical Chemistry,* 5th ed., Allyn and Bacon (Boston), 1987.

3. David R. Lide, Editor-in-Chief, *CRC Handbook of Chemistry and Physics*, 77th ed., CRC Press (Boca Raton, FL), 1996.

4. D. A. Skoog, D. M. West, and F. J. Holler, *Fundamentals of Analytical Chemistry,* 7th ed., Saunders (New York), 1996.

5. For more about the Q-test, refer to References 2 and 4.

Experiment 2

Collection and Preservation of Water and Sediment Samples: Inorganic and Organic Profiles of Soil and Sediment Cores

Objectives—The objectives of this experiment are:

1. To introduce one of the most important aspects of environmental science, sampling, and sample preservation

2. To apply the concepts of sampling to the collection of a water sample and a sediment (or soil) core

3. To test the core for its gross inorganic and organic content

Introduction—Scientific methods have been developed for sampling solids, liquids, and gases; however, very little time is spent in teaching the fundamentals of sampling. A search in the *Journal of Analytical Chemistry* over a random 6-month period located a total of 68 articles published on various methods of chemical analysis, but only six had any mention of collecting samples for analysis. Two important references to the practice and theory of sampling are given in the literature section at the end of this experiment.

Theory

Sampling Error Time should be spent designing a valid sampling protocol. Any step in the procedure, beginning with removing the sample from the bulk matrix, can introduce error; however, an error in sampling will produce uncertainty in any result. When there is more than one source of random error, the total standard deviation of a result (s_o) is the square root of the sum of the squares of the standard deviations of each measurement or step. Lumping all errors into two classes, "sampling" and "analysis," one can get an appreciation for how exact an analytical procedure should be for a particular sampling method:

$$s_o = [\,(s_s)^2 + (s_a)^2\,]^{1/2} \qquad (2\text{-}1)$$

where s_o is the overall standard deviation, s_s is the standard deviation of sampling, and s_a is the standard deviation of the analysis.

Using Eq. 2-1, it can be shown that if the standard deviation of the analysis is 30% of the standard deviation of sampling, only 4% of the overall standard deviation is due to the analysis. Thus, the overall accuracy required of an analysis need be no better than the minimum required from the sampling. As a rule of thumb, when s_a is less than one-third of s_s further improvement in analytical accuracy is wasted. Equation 2-1 also confirms that the accuracy of a result may be determined by the sample collection method used.

In-Situ Analysis One way to avoid sampling error is to avoid sampling altogether. An on-site analysis, if practical, is the preferred approach. The drawback is that a subsequent analysis is not possible unless an actual sample is removed. Many in-situ monitors exist which make the analytical measurement in the bulk material. These include (1) instruments for pH, temperature, turbidity, conductivity, chemical oxygen demand, (2) devices specific for nitrogen, phosphorus, cyanide, etc., and (3) ion selective electrodes for species such as chloride, fluoride, nitrate, potassium, and many others.

Sampling Plan The use of a carefully thought-out sampling plan minimizes the sampling error. The plan should include site-selection criteria, methods and equipment for collecting the sample, proper storage of the sample prior to analysis, and the ultimate disposal of the sample after completing the analysis (including archival storage, if appropriate).

Sample Storage The type of sample and the desired analyte dictate the proper storage bottle (glass, polyethylene, Teflon, or other materials), the storage method, and the maximum storage time. The addition of a "spike" of a chemically-similar nonanalyte at known concentration at the time of sampling permits an estimation of the deterioration of a sample on storage (if the spike has not decreased in concentration, the analytes may not have either). Table 2-1 describes some recommendations for sample storage.

Part A: Water

Two types of water samplers are available—grab and automated. A grab sampler is, as its name indicates, a crude mechanism for collecting a sample. This sampler can be as simple as a bucket lowered over the side of a ship, a bottle lowered to a depth and the lid removed, or a container dropped on a weighted tether with an activator (e.g., a weight dropped down the line) which opens a port or ports (e.g., Niskin Bottle, or Kemmerer Sampler, shown in Figure 2-1). The type of sampler used is determined by the type of analysis to be performed.

Some general rules for sampling a body of water include:

a. Collect the sample as close to the target-type as possible (at a water intake, a sewage outfall, etc.).
b. If possible, sample a moving stream (if appropriate).
c. Sample in the middle of a lake or river (however, few pollutants are usually found here).
d. Sample at mid-depth if a representative sample of the bulk water is desired.
e. Sample no sooner than necessary to avoid aging of the sample.

Table 2-1 EPA Recommended Preservation Methods of Water and Wastewater Samples

Parameter	Preservation Method	Container*	Maximum Holding Time
Acidity/Alkalinity	Store at 4°C	P, G	14 days
Ammonia	Sulfuric acid to pH < 2	P, G	28 days
	Store at 4°C		
BOD	Store at 4°C	P, G	48 hours
COD	Sulfuric acid to pH < 2	P, G	28 days
	Store at 4°C		
Chloride	None	P, G	28 days
Residual chlorine	None	P, G	Analyze immediately
	0.6 g Ascorbic acid		
Dissolved oxygen	None	G with Glass Top	Analyze immediately
Fluoride	None	P	28 days
Mercury	Nitric acid to pH < 2	P, G	6 months
Nitrate	Sulfuric acid to pH < 2	P, G	48 hours
	Store at 4°C		
Nitrite	Store at 4°C	P, G	48 hours
Oil and grease	Sulfuric acid to pH < 2	G	28 days
	Store at 4°C		
Total organic carbon	Sulfuric acid to pH < 2	P, G	28 days
	Store at 4°C		
pH	None	P, G	Analyze immediately
Ortho-Phosphate	Filter on site	P, G	48 hours
	Store at 4°C		
Phosphorus, total	Sulfuric acid to pH < 2	P, G	28 days
Solids	Store at 4°C	P, G	7 days
Specific conductance	Store at 4°C	P, G	28 days
Sulfate	Store at 4°C	P, G	28 days
Sulfide	Store at 4°C	P, G	7 days
Turbidity	Store at 4°C	P, G	48 hours
Purgeable aromatic Hydrocarbons	Store at 4°C	G, Teflon-Lined Septum	14 days
	HCl to pH 2		
Phenols	Store at 4°C	G, Teflon-Lined Cap	7 Days until extraction
			40 Days after extraction
PCBs	Store at 4°C	G, Teflon-Lined Cap	Same as above
Phthalate esters	Store at 4°C	G, Teflon-Lined Cap	Same as above
	Store in dark		

*P is plastic and G is glass.

Source: EPA-600/4-82-0129, Handbook for Sampling and Sample Preservation in Various Waters and Wastewaters.

Figure 2-1 Basic Kemmerer-Type Water Sampler

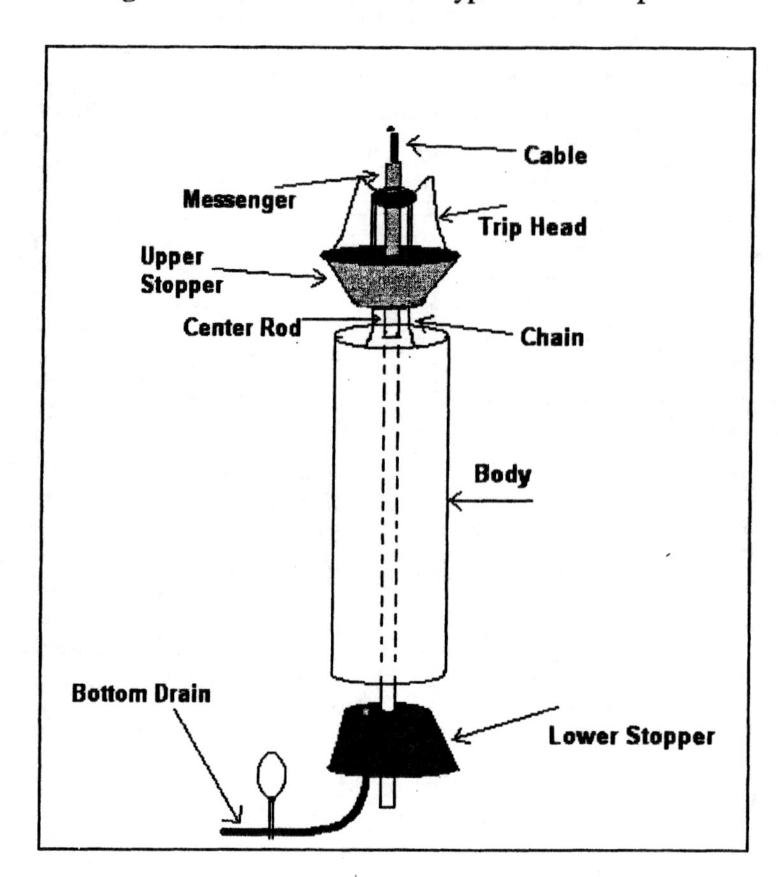

Complicating Factors in Sampling Include:

a. The presence of sediment which may settle out, carrying (or releasing) analyte with it

b. Biological life in the sample, which may metabolize sample components

c. The presence of gases, which may desorb when brought from depth (release of pressure)

Many other situations, not imagined in the laboratory, may require attention at the time of sampling (e.g., the spraying of insect repellents may contaminate the sample if it is not enclosed).

Automated samplers are well suited for flowing water or process streams (pipe, outfall, etc.). These devices can be set to sample a set fraction of the total flow of a stream or remove a subsample at fixed time intervals (which may or may not be combined, as the sampling plan requires). These devices are not suitable for sampling streams with high solids content, which may cause clogging.

Part B: Sediment

The type of sediment sample collected depends on the objective of the project. Bottom samples of two types are commonly obtained: grab samples, which take a portion of the uppermost layers of the sediment and core samples that penetrate the sediment to some depth. (For details, see Reference 3 in the Literature section.)

Ponar Grabs A Ponar sampler is used to collect the upper 10 cm of sediment. Ponar samples are usually collected from shipboard. The sampler, shown in Figure 2-2, is lowered on a line to the bottom. It consists of a pair of weighted jaws on a hinged apparatus that closes when the sampler hits the bottom (or when a "messenger" weight is sent down the line).

Figure 2-2 A Sediment Grab Sampler

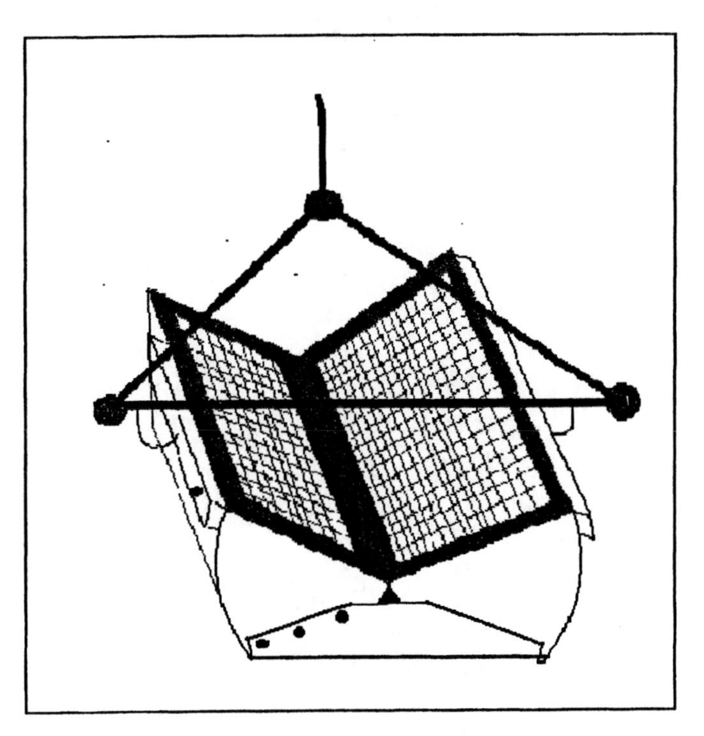

The sampler is covered with a screen (to keep the sample in) and a rubber flap (to let the water out). Both reduce sample loss as the sampler is brought to the surface. If a bulk sample is required, the Ponar Grab can be opened over a large container and the entire contents saved (about 20 kg of sediment); alternatively, a representative sample from the surface sediment can be removed from the bulk sample.

Coring Core samples provide a vertical profile of the sediment at the bottom of a stream, lake, pond, or ocean. For undisturbed sediments, the topmost layer is the most recent—the deeper the layer, the longer ago it was deposited. A core is essentially an elapsed time indicator of processes that occurred in the past.

A simple and inexpensive coring device can be constructed from materials obtainable from a hardware store; more complicated devices contain plastic or metal sleeves, steel casings, and a "core catcher"(to prevent the sediment from slipping out.) One simple "corer" is shown in Figure 2-3.

Safety Issues

1. Safety glasses must be worn at all times in a chemistry laboratory.

2. When collecting samples from a stream, lake, or other body of water, life vests must be worn. Also, the instructor must be present when collecting samples from the environment.

3. When collecting samples from a boat, a qualified person must be in charge of the boat.

Figure 2-3 Schematic Diagram of the Sampling Device Used to Collect Sediment Cores (Components Are Made from 1- Inch - Diameter Schedule 40 PVC)

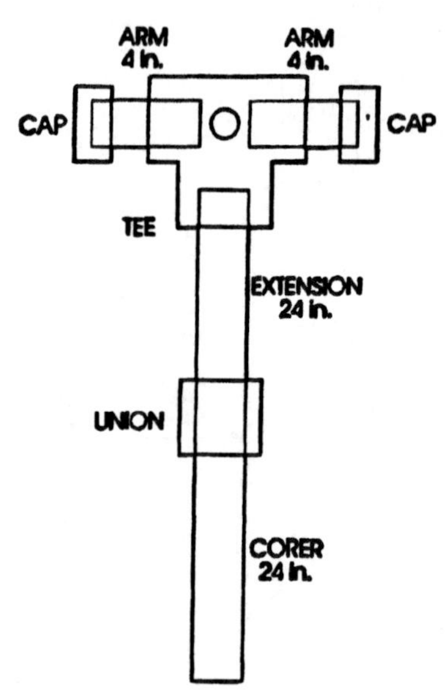

Source: R. D. Delumyea and D. L. McCleary, A device to collect sediment cores, J. Chem. Ed., 70, 1993, 172-173. By permission.

Procedure

Part A: Collecting a Water Sample

1. Ideally, the student should gain experience using various types of devices for collecting water samples. In the first experiment one type of device was used (shown in Figure 1-5), and in this experiment a second type will be used. If the sample is collected from a boat, it will be a grab sample of the surface water. A grab sample at a known depth can be obtained by using a weighted bottle on a string. A widely used grab sampler is the Kemmerer bottle, shown in Figure 2-1, which consists of a spring-loaded stopper at each end. A "messenger" closes both stoppers, trapping the sample in the PVC tube.

2. It is good practice to store samples in ice until returning to the laboratory. Collect a 1 L water sample using the device provided by the instructor. Note all sampling conditions and clearly identify the bottle with a mark or label.

3. On returning to the laboratory, the bottle should be refrigerated at 4°C. A label giving the collection location, time, date, and name of the collector(s) should be put on the bottle. This sample will not be studied in the present experiment, but will be used in several subsequent experiments.

Part B: Collecting a Grab and/or Core Sample

A major goal of this experiment, in addition to learning sampling methods, is to obtain a core sample, section it, and treat the sections to determine the organic and inorganic contents. The segments can be used at a later time for metals analysis using colorimetry or atomic absorption methods.

1. If a Ponar Grab is available, it is used on shipboard to collect a sediment sample that will be analyzed for several metals. The depth at which the sample was collected should be recorded as well as collection conditions. When retrieved, the sampler is opened above a large aluminum tray and emptied into it. The sediment can then be scooped into jars (Mason jars are convenient), which are kept on ice until returning to the lab. The jars are stored in a refrigerator at 4°C.

2. Core samples are difficult to collect from the side of a ship. The exact details for collecting a core depend on the nature of the sampling site. For simplicity, the following procedure will be given that applies to shallow waters with an approachable shore. In practice, it is a good idea to take preliminary cores to determine if the site is suitable.

To use the coring device of Figure 2-3, the core must be sufficiently rigid to support itself when extruded. Very wet or sandy (i.e., fluid) sediments will run out when being removed from the corer; in that case, a core catcher is needed.

The core is collected by assembling the device shown in Figure 2-3 and, holding it vertically, slowly pressing from above without twisting until the corer reaches a stopping point. The corer is withdrawn by pulling up, and twisting, if necessary. A stopper is inserted into the lower end of the corer and the extension is removed. Another stopper is inserted at the top of the corer, which is stored vertically until the core is sectioned.

Part C: Sectioning the Core

1. Once a core has been obtained, it is taken to the laboratory to be sectioned into subsamples for analysis. The extruder is carefully inserted at the base of the corer and pushed up until the sample is even with the top of the tube. At this point, some sample loss may occur if water was collected, or if the sample is particularly soupy. The extruder is marked from this point at intervals determined by the experimenter—2 cm is recommended. The length of the section determines the resolution of the profile; a smaller section reduces the amount of sample available for analysis.

2. The extruder is pushed to the first mark and the section cut off using a flat spatula or a knife. The section is collected in a preweighed (to 0.1 mg) porcelain evaporating dish (60 mm diameter) to air-dry overnight. Make sure the dishes are marked for identification.

Part D: The Organic and Inorganic Profile of the Core

1. Weigh (to 0.1 mg) the air-dried sections and then place the dishes in a drying oven at 110°C for 3 hours.

2. Remove the dishes from the oven, allow them to cool, and then reweigh.

3. The dried segments are then ashed with either a Meker burner (which takes about 45 minutes) or in a muffle oven at 600°C (this is the preferred procedure) for one hour. This results in the combustion of soil organic matter, which is lost as carbon dioxide and water. After removing the samples from the muffle oven, allow them to cool and then weigh.

4. Transfer the ashed sections to bottles or vials where they are stored until a later time when metals analyses of the samples may be carried out.

Figure 2-4 shows the organic profile of a sample taken from a river in the authors' area. In general, there is a direct relationship between the amount of organic matter and the depth of the sediment down to the beginning of the sand layer (approximately 13 cm for the given sample).

Figure 2-4 Percentage Organic Matter at Various Depths in a Sediment Core

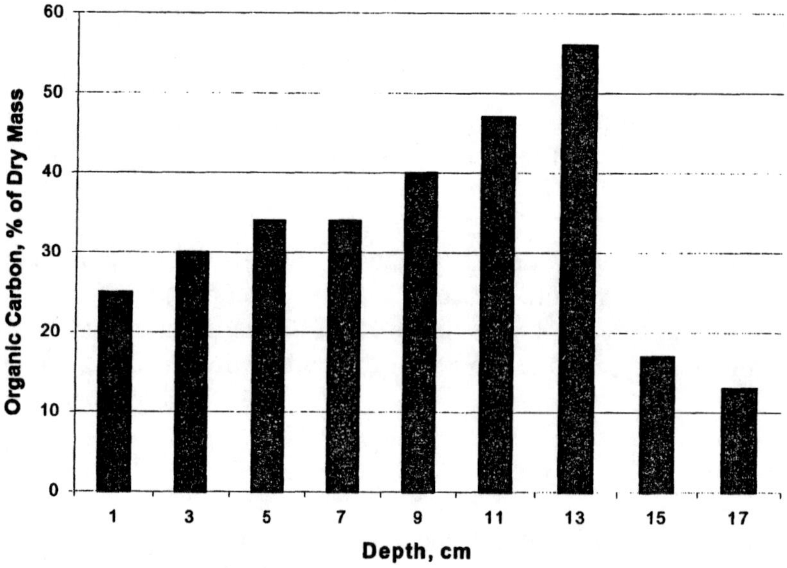

Source: R. D. Delumyea and D. L. McCleary, A device to collect sediment cores, J. Chem. Ed., 70, 1993, 172-173. By permission.

Waste Minimization and Disposal

1. Since samples are often collected from areas or under conditions where we expect to see some substantial levels of pollutants, the water and sediments must be considered to be hazardous wastes. When the samples are no longer needed, they cannot be rinsed down the drain or disposed of in the trash, and they cannot be taken back to the sampling site. Your instructor will give directions for collecting the waste and will see to their proper disposal.

2. The sediment and water samples may be used in several subsequent experiments where pollutants may be destroyed or inactivated (by precipitation of metals, for example). To facilitate and lessen the cost of disposal, a minimum amount of water and sediment should be collected. One liter of water is sufficient for many tests, and 200 g of sediment can be used for many experiments.

Data Analysis

1. Submit a data sheet similar to the one shown below that contains all of the experimental and calculated data. The data sheet shows the organization of the pertinent quantities. Heating in the muffle oven burns off the organic matter and leaves inorganic matter behind. The difference in mass before and after firing in the muffle oven gives the mass of organic material. The mass of inorganic material is simply the difference between the final mass and the mass of the empty dish.

DATA SHEET

Table of Mass Measurements Used to Determine the Organic/Inorganic Profile of a Core Sample

Section	Depth(cm)	Mass of Dish(g)	Dry Weight(g)	Ash Weight(g)
1	0–2			
2	2–4			
3	4–6			
4	6–8			
5	8–10			
6	10–12			
7	12–14			
8	14–16			

2. Prepare a plot like that shown in Figure 2-4 using your results (generating the plot with computer software, if possible).

3. Report the nature of the core sections (sandy, clay-like, mucky, etc.), and relate these features to the calculated organic content. High inorganic content is expected if the sample is mainly sandy, and a silty matrix is often high in organic matter. Some core sections are colored, and the color of each section should be recorded in your laboratory notebook.

4. Report the weather and water conditions when the water, sediment, and core samples were collected.

5. Be sure to pinpoint the exact location of the sampling site!

6. The difference between the original sample mass and its mass after drying in a drying oven can be used to convert wet masses into dry masses. This is very useful in subsequent analyses that may require the dry mass of sample. Calculate the ratio of dry mass to wet mass and report the values obtained for your samples.

Supplemental Activity

1. It would be interesting for students to take cores from different locations so sample comparisons can be made. Also, a soil sample is different from a sediment taken from a body of water.

2. Each section of core can be examined for metals, such as iron or manganese, and a profile obtained for

the metal. This can be done colorimetrically (See Experiment 9) or by atomic absorption (See Experiment 10).

Questions and Further Thoughts

1. In an estuary fresh water meets salt water, which serves to coagulate colloidal particulates, resulting in considerable silting. This is one reason for the formation of a delta at the mouth of a river into the ocean. Such silt contains many of the impurities that were originally present in the water. Thus, silt makes good samples to study.

2. What assumptions are made in this experiment? What errors result from the presence of volatile or unstable inorganic nitrogen compounds in the sample?

Notes

1. When collecting samples, especially aboard ship, a lot of activity occurs. In addition to collecting samples, students must record water and air temperatures, locate landmarks, record water depth, measure salinity, label bottles, and carry out many other activities. Because of this activity and the movement of the boat and the messiness of the operation, it is difficult to make entries in a lab notebook. It is much more convenient to use a tape recorder to take notes and later transcribe them into a lab notebook.

2. A muffle oven takes a long time to cool, and it saves time to let samples cool in the oven overnight.

Literature Cited

1. Sampling-related topics and analysis: A. E. Greenberg, L. S. Clesceri and A. D. Eaton, *Standard Methods for the Examination of Water and Wastewater,* 18th ed., American Public Health Association, Washington, DC, 1992.

2. Concepts underlying sampling theory: B. Kratochvil and J. K. Taylor, Sampling for chemical analysis, *Anal. Chem., 53(8)*:924A–938A, (1981).

3. Devices and potential problems of sediment sampling: A. Murdoch and S. D. MacKnight, Eds., *CRC Handbook of Techniques for Aquatic Sediments Sampling,* CRC Press, Boca Raton, FL, 1991.

4. EPA-600/4-82-029, *Handbook for Sampling and Sample Preservation in Various Waters and Wastewaters, Cincinnati, O, 1982.*

Inorganic Chemical Properties of Natural Waters and Wastewaters

2

Experiment 3

The pH and Buffer Capacity of Environmental Waters

Objectives—The objectives of this experiment are:

1. To learn the use of the pH electrode by making pH measurements on environmental waters, rainwater and household products

2. To determine the buffer capacity of a natural water

Introduction— Natural waters contain a wide variety of organic and inorganic solutes. Some of the more important solutes are inorganic acids and bases and organic acids and bases. The inorganic acids include sulfuric and nitric acids from both anthropogenic and natural origins. The organic acids include humic and tannic acids, which result from the decomposition of organic matter. Carbon dioxide from the atmosphere contributes significantly to the acidity of environmental waters due to its conversion to carbonic acid, H_2CO_3.

Acids may be defined as substances that increase the concentration of $H^+(aq)$ in water, according to the most elementary view of acids, that of Arrhenius. A broader view, that of Brønsted and Lowry, defines an acid as a proton donor, illustrated by the equation,

$$\mathbf{HCl(g) + H_2O(aq) \longrightarrow H_3O^+(aq) + Cl^-(aq)}$$

where $H_3O^+(aq)$ is chemically equivalent to $H^+(aq)$. An even broader definition of an acid, that of G. N. Lewis, defines an acid as an electron-pair acceptor, as in

$$\mathbf{H^+(aq) + :OH^-(aq) \longrightarrow H—O—H(l)}$$

or

$$\mathbf{Fe^{3+}(aq) + :SCN^-(aq) \longrightarrow Fe—SCN^{2+}(aq)}$$

Bases, as defined by Arrhenius, are substances that increase the hydroxide ion concentration in water. Thus, KOH and NaOH are bases as well as the weak base, ammonia:

$$NH_3(g) + H_2O(l) \longrightarrow NH_4^+(aq) + OH^-(aq)$$

As with acids, broader and more useful definitions of bases can be devised. Brønsted and Lowry defined a base as a proton acceptor, illustrated by the previous equation. The Lewis theory defines a base as an electron-pair donor, which again makes ammonia a base.

The concentration of hydronium ion found in the laboratory is highly variable, with values of less than 10^{-13} M observed for very basic solutions and values greater than 10 M observed for concentrated acids.

To simplify expressing such disparate concentrations, the logarithm to the base 10 is used to convert hydrogen ion concentration, in moles per liter $[H^+]$, into what is termed the pH. By definition then,

$$pH = -\log [H^+] \tag{3-1}$$

where the negative sign is used to make most pH values positive. A more exact definition of pH uses the activity of the hydronium ion rather than its actual molar concentration. The activity is the "effective" concentration. However, for dilute solutions the difference between activity and actual molarity is small enough to ignore.

The pH of neutral water at 25°C is 7.00, whereas acid solutions have a pH less than 7 and basic solutions have a pH greater than 7. Neutrality is defined as the condition where $[H^+] = [OH^-]$. The approximate pH values of several substances are given in Table 3-1. The pH of pure water is temperature dependent, decreasing with increasing temperature since the dissociation of water increases with increasing temperature (dissociation is endothermic).

Table 3-1 The pH of Various Fluids

System	pH	System	pH
Stomach acid	1.7	Beer	4.0–5.0
Lemon juice	2.2–2.4	Cow's milk	6.3–6.6
Vinegar	3.0	Potable water	6.5–8.0
Tomato juice	3.5	Human blood	7.3–7.5
Wine	3.6	Human urine	4.8–8.4

Theory

1. **The Origin of Natural Acidity.** In this section we examine factors that affect the pH of natural waters, which are often somewhat basic. Dissolved salts such as sodium chloride do not contribute to the acid or basic nature of environmental waters. Some salts, such as ferric chloride, do affect pH due to hydrolysis,

$$Fe^{3+}(aq) + H_2O(l) \rightleftharpoons [Fe(OH)]^{2+}(aq) + H^+(aq)$$

where Fe^{3+} acts as a Lewis acid. Small, highly charged ions behave in this manner, metal ions producing acid solutions and anions, such as fluoride, producing basic solutions,

$$F^-(aq) + H_2O(l) \rightleftarrows HF(aq) + OH^-(aq)$$

The fluoride ion acts both as a Lewis and a Brønsted-Lowry base in this example. The reaction proceeds to the right to some extent since HF is a weak acid.

Most acidity in natural waters is due to carbon dioxide, which, when dissolved in water, produces hydronium ion:

$$CO_2(g) + H_2O(l) \rightleftarrows H^+(aq) + HCO_3^-(aq)$$

This reaction can be considered to occur in two stages. **Stage I** is the establishment of equilibrium between atmospheric and aqueous carbon dioxide. The amount of carbon dioxide that dissolves in water is governed by Henry's law, which takes the form

$$[CO_2(aq)] = K_H P(CO_2) \qquad (3\text{-}2)$$

where K_H is the Henry's law constant and $P(CO_2)$ is the partial pressure of carbon dioxide.

For carbon dioxide at 25°C, $K_H = 3.4 \times 10^{-2}$ mol/L-atm. The concentration of CO_2 in the atmosphere is about 350 ppm. In the gaseous state, 350 ppm of CO_2 means 350 molecules per 1×10^6 molecules of air, and since moles and molecules are proportional,

$$350/1000000 \cong \text{mol } CO_2 / \text{mol air} \cong \text{mol fraction } CO_2$$

$$= \text{pressure fraction } CO_2 = P(CO_2)/P(\text{air})$$

Thus $P(CO_2) = 3.5 \times 10^{-4}$ atm when $P_{air} = 1.0$ atm and $[CO_2(aq)] = 1.2 \times 10^{-5}$ mol/L at 25°C from Stage I.

Stage II is the dissociation of dissolved carbonic acid, according to

$$H_2CO_3(aq) \rightleftarrows H^+(aq) + HCO^-(aq) \qquad K_{a1} = 4.5 \times 10^{-7}$$

$$HCO_3^-(aq) \rightleftarrows H^+(aq) + CO_3^{2-}(aq) \qquad K_{a2} = 4.7 \times 10^{-11}$$

Since $K_{a1} \gg K_{a2}$, the pH of the system is primarily due to the first equilibrium. Therefore, $K_{a1} = [H^+][HCO^-]/[H_2CO_3] = [H^+]^2/1.2 \times 10^{-5} = 4.5 \times 10^{-7}$, and $[H^+] = 2.3 \times 10^{-6}$, giving a pH of 5.63. **This is the expected** pH for pure water in equilibrium with atmospheric carbon dioxide at 25°C.

Although the pH of a natural water is affected by the carbon dioxide acidity of rain, a third factor to be considered is the background carbonate level that is due, in part, to the dissociation of calcium carbonate in soil. When rain falls on land, it first percolates through topsoil, where its pH may drop by another unit due to the large quantity of carbon dioxide produced by bacteria. However, much of the earth's crust contains calcium carbonate (ultimately derived from marine organisms).

The effect of calcium carbonate on the pH of environmental waters is due to three factors: (1) calcium carbonate is sparingly soluble in water; (2) the carbonate ion is a moderately strong base, whereas bicarbonate is only a weak base, and (3) dissolved carbonate is in equilibrium with carbon dioxide in soil gases and in bodies of water. The net result of these factors is that the pH of natural waters will be somewhat basic instead of the acidic pH from dissolved CO_2 alone. To illustrate, first consider the dissociation of calcium carbonate,

$$\mathbf{CaCO_3(s) \rightleftarrows Ca^{2+}(aq) + CO_3^{2-}(aq)}$$

where $K_{sp} = [Ca^{2+}][CO_3^{2-}] = 4.6 \times 10^{-9}$ at 25° C. If this were the only process to occur, the **pH would remain unchanged.**

However, since carbonate ion is a Brønsted-Lowry base, according to

$$\mathbf{CO_3^{2-}(aq) + H_2O(l) \rightleftarrows HCO_3^-(aq) + OH^-(aq)}$$

the observed pH is due to not only atmospheric carbon dioxide, but also to the carbonate from the hydrolysis of $CaCO_3(s)$, which increases pH.

The hydrolysis constant for the last reaction is found as follows:

$$\mathbf{K_h = K_w / K_{a2} = [HCO_3^-][OH^-]/[CO_3^{2-}]}$$

$$\mathbf{= 1.0 \times 10^{-14} / 4.7 \times 10^{-11} = 2.1 \times 10^{-4}}$$

When the solubility product equilibrium is combined with the hydrolysis equilibrium, the net result is

$$\mathbf{CaCO_3(s) + H_2O(l) \rightleftarrows Ca^{2+}(aq) + HCO_3^-(aq) + OH^-(aq)}$$

since when equilibria are added the new equilibrium constant is the product of the individual equilibrium constants, the equilibrium constant for the last equilibrium is $K' = K_{sp} \times K_h = 4.6 \times 10^{-9} \times 2.1 \times 10^{-4} = 9.7 \times 10^{-13}$. Since $[Ca^{2+}][HCO_3^-][OH^-] = [OH^-]^3 = 9.7 \times 10^{-13}$, $[OH^-] = 9.9 \times 10^{-5}$ and pH = 10.00. Thus, without atmospheric carbon dioxide, the pH of natural waters in contact with calcium carbonate would be quite high.

Now consider both equilibria simultaneously,

$$\mathbf{H_2CO_3(aq) \rightleftarrows H^+(aq) + HCO_3^-(aq)}$$

$$\mathbf{CaCO_3(s) + H_2O(l) \rightleftarrows Ca^{2+}(aq) + HCO_3^-(aq) + OH^-(aq)}$$

and include the equilibrium that we know also occurs simultaneously,

$$CaCO_3(s) + H_2O(l) \rightleftharpoons Ca^{2+}(aq) + HCO_3^-(aq) + OH^-(aq)$$

and include the equilibrium that we know also occurs simultaneously,

$$H^+(aq) + OH^-(aq) \rightleftharpoons H_2O(l)$$

where K_{a1} = 4.5×10^{-7}, K' = 9.7×10^{-13}, and $K = 1/K_w$ = 1.0×10^{14}, respectively for the three equilibria. The overall description of the three simultaneous processes is found by summing the equilibria to give,

$$CaCO_3(s) + H_2CO_3(aq) \rightleftharpoons 2HCO_3^-(aq) + Ca^{2+}(aq)$$

for which K'' = $K_{a1} \times K' \times K = 4.4 \times 10^{-5}$.

Now, since $[HCO_3^-] = 2 \times [Ca^{2+}]$, $[Ca^{2+}](2 \times [Ca^{2+}])^2 / 1.2 \times 10^{-5} = 4.4 \times 10^{-5}$, then $[Ca^{2+}] = 5.1 \times 10^{-4}$ mol/L. In terms of ppm $CaCO_3$, the calcium level is (5.1×10^{-4} mol/L)(100 g/mol)(1000mg/g) = 51 ppm—a reasonable value based on actual levels found.

Finally, to calculate the expected pH, the last equilibrium is used and we examine the bicarbonate produced to see if it is a stronger acid or a stronger base. As an acid,

$$HCO_3^-(aq) \rightleftharpoons H^+(aq) + CO_3^{2-}(aq) \qquad K_{a2} = 4.7 \times 10^{-11}$$

and as a base,

$$HCO_3^-(aq) + H_2O(l) \rightleftharpoons H_2CO_3(aq) + OH^-(aq) \qquad K_b = 2.2 \times 10^{-8}$$

since K_b = K_h = K_w / K_{a1}. Since $K_h \gg K_{a2}$, K_{a2} can be ignored and the pH is calculated using the expression for K_b.

Then, K_b = $[H_2CO_3][OH^-]/[HCO_3^-]$ = 2.2×10^{-8}. From Henry's law we found that $[H_2CO_3]$ = 1.2×10^{-5} and we also showed that $[HCO_3^-]$ = $2 \times [Ca^{2+}]$ = $2(5.1 \times 10^{-4})$ = 1.0×10^{-3}. Since K_b is so small, we can use $(1.2 \times 10^{-5})[OH^-] / 1.0 \times 10^{-3} = 2.2 \times 10^{-8}$. This gives $[OH^-] = 1.8 \times 10^{-6}$, and finally pH equals 8.26. This is remarkably close to the pH of many natural waters. Over a 10-year period in a major stream in the southeastern United States, the authors found an average pH close to 8.0, with little variation, except after periods of substantial rainfall.

2. **Buffers and Buffer Capacity**. We have seen that natural waters can contain appreciable quantities of bicarbonate ion and dissolved carbon dioxide. Thus, such waters can neutralize both strong acids and bases, according to:

$$HCO_3^-(aq) + HCl(aq) \longrightarrow H_2O(l) + CO_2(g) + Cl^-(aq)$$

and

$$H_2CO_3(aq) + NaOH(aq) \longrightarrow H_2O(l) + Na^+(aq) + HCO_3^-(aq)$$

Therefore natural waters act as buffers—that is, solutions that resist pH changes when strong acids or bases are added.

Buffer solutions contain a weak acid (H_2CO_3 as shown) and a salt of that weak acid ($NaHCO_3$ as shown), or a weak base and a salt of the weak base (ammonia and ammonium chloride as an example). This assumes comparable concentrations of acid and base. In nature, acid buffers are much more common, and we shall consider only this type.

To appreciate the effect of buffering action, consider the effect of one drop (about 0.050 mL) of concentrated HCl (12 M) when added to one liter of water. The hydronium ion concentration is (0.050 mL/1000 mL/L)(12 mol/L) /1.0 L = 6.0×10^{-4} M, giving a pH of 3.22, a decrease of about 4 units!

If a strong base is added to pure water, say one pellet of about 0.10 g of NaOH, the hydroxide concentration will be (0.10 g/40 g/mol) / 1.0 L = 0.0025 M, giving a pH of 11.40, an increase of more than 4 units. Typical buffers change less than ±0.1 unit upon the addition of small amounts of strong acid or base.

To understand how to make buffer solutions of a given pH and to calculate the pH of specific buffers, we shall consider an acid buffer made by mixing a generic weak acid, HA, with a salt of the acid, NaA, for example. The dissociation of the acid is used to develop the necessary equations:

$$HA(aq) \rightleftharpoons H^+(aq) + A^-(aq)$$

and

$$K_a = [H^+][A^-] / [HA] \quad (3\text{-}3)$$

Rearranging Eq. 3-3 gives

$$[H^+] = ([HA] / [A^-])K_a \quad (3\text{-}4)$$

and taking the negative of the logarithm to the base 10 of both sides gives

$$pH = pK_a + \log([A^-] / [HA]) \quad (3\text{-}5)$$

where $pK_a = -\log(K_a)$ and $[A^-] / [HA]$ is the ratio of the concentrations of the conjugate base to the acid.

A buffer of specified pH can be made by adjusting the ratio of base to conjugate acid. Likewise, for a given base-to-acid ratio, the pH of the buffer system can be calculated. In the special case when acid and base concentrations are equal, pH = pK_a and the buffer is said to be centered at "pK_a." For acetic acid, with $K_a = 1.8 \times 10^{-5}$, the buffer is centered at $-\log 1.8 \times 10^{-5} = 4.74$. Generally, buffered systems are made to be as closely centered as possible.

We now examine what happens to the pH when one drop of 12 M HCl is added to 1 L of the previous buffer. Various amounts of acid and conjugate base may be used at the buffer center (or any other pH value), but the capacity of the buffer to resist pH changes is affected by the quantities of acid and conjugate base present. Suppose that, in our example, 0.100 mol of acetic acid, HAc, and 0.100 mol of sodium acetate, NaAc, are present in 1.00 L of solution. When 1 drop of the acid (0.60 mmol) is added, it reacts with an equivalent amount of base, according to $Ac^- + H^+ \rightarrow HAc$. This leaves $1.00 \times 10^2 - 0.60 = 99$ mmol of Ac^-, and results in a total of $1.00 \times 10^2 + 0.6$ mmol = 101 mmol of HAc. Thus, according to Eq. 3-5 the final pH is $\log(99/101) - \log(1.8 \times 10^{-5}) = 4.74$. Thus, no significant change in pH occurs. A similar result occurs when a small amount of strong base is added.

Since **buffer capacity is defined as the moles of strong acid (or strong base) needed to change the pH of 1.00 L of a buffer by 1 pH unit**, we can use the previous buffer system to show how this quantity is determined. The buffer capacity is the amount of HCl, for example, needed to change the pH of the prior buffer system from its center at 4.74 (the buffer center is used for convenience) to a final value of 3.74. Then, $3.74 = \log([A^-])/[HA]) + 4.74$, or $[HA]/[A^-] = 1.0 \times 10$. Let y = mol of HCl added. Thus, from $H^+ + A^- \rightarrow HA$, $y = [H^+]$, $[A^-]_0 = 0.10$ and the final value of $[HA] = 0.10 + y$. At equilibrium, $[A^-] = 0.10 - y$, and thus $[HA]/[A^-] = (0.10 + y)/(0.10 - y) = 1.0 \times 10$, which finally gives $y = 0.082$ mol. Thus the buffer capacity of 0.10 M acetic acid buffer centered at 4.74 is 82 mmol. For the 12 M HCl solution we found that 1 drop was equivalent to about 0.60 mmol, and therefore 137 drops of 12 M HCl would change the pH of the buffer by only about 1 unit.

3. **Ion-Selective Electrodes**—The Glass Electrode. A transducer converts information from the chemical domain into the electrical domain. A simple example of a transducer is an ion-selective electrode (ISE), also called a specific ion electrode. These electrodes can detect very low concentrations of certain ions in the presence of other ions (although interferences can occur). The electrode develops a potential that is related to the concentration (or activity) of the ion, and the potential is measured with a potentiometer. Many ion-selective electrodes have been developed in the last 30 years and are finding uses in pollution studies, industrial system monitoring, medical applications, and many other areas.

The familiar glass electrode, used to measure hydrogen ion activity, is an example of an ISE. It is very selective for hydronium ion, a result of the composition of the glass membrane, the glass wall in contact with the test solution. In fact, all ISEs are membrane electrodes with a thin section separating two solutions across which an electrical potential is developed.

An ISE consists of a glass (or plastic, epoxy, etc.) tube sealed at one end with the membrane and contains a solution of the ion being determined. Electrical contact is made to the filling solution with a calomel or silver/silver chloride internal reference electrode. When the electrode is placed in a sample solution, the sensing membrane allows only the ion of interest to "migrate," that is, there is a momentary flow of analyte ions in the membrane in the direction of the solution containing the smaller concentration of that ion. Since ions carry a charge, an electrical potential is produced that opposes further ion displacement, and an equilibrium is established. A typical pH combination electrode (one containing an internal reference electrode) is illustrated in Figure 3-1.

Figure 3-1 Combination Glass Electrode

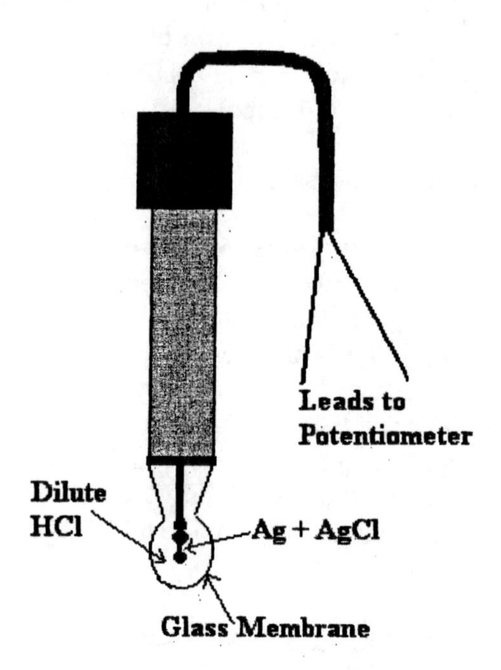

The potential of an ISE is expressed mathematically by the Nernst equation,

$$E_{\text{ind el}} = E^\circ + (RT/nF) \ln (a_x^{n+}) \tag{3-6}$$

where the ion under study is a cation of charge *n*+. In this equation *R* is the gas constant, *T* is the absolute temperature, *F* is the Faraday constant (96,485 C/mol), and *a* is the activity of the ion of interest. The potential of the sensing or indicator electrode, is $E_{\text{ind el}}$. For the pH electrode at 25°C,

$$E_{\text{ind el}} = E^\circ_{H+} + 0.05915 \log(a_{H+}) = -0.05915 \text{ pH} \tag{3-7}$$

since the electrode potential for a standard hydrogen electrode is 0.00 V.

Potentiometric measurements are made with an electrochemical cell composed of two electrodes, the indicating and the reference electrodes. The indicating electrode responds to changes in the concentration of the chemical species to be determined, according to the Nernst equation. The reference electrode is self-contained and has a constant potential. It is equipped with a fiber tip that provides electrical contact with the sample solution.

The measured potential is

$$E_{\text{meas}} = E_{\text{ind el}} - E_{\text{ref}} \tag{3-8}$$

For the pH electrode this equation becomes

$$E_{meas} = C - 0.0591\text{ pH} \qquad \textbf{(3-9)}$$

where C is a potential due to the reference electrode.

Before using a pH electrode, it must be calibrated with buffer solutions of known pH. From Eq. 3-9 it is seen that a plot of measured potential versus pH should give a straight line with slope equal to –0.05915. In a pH meter (one type of potentiometer) calibration establishes a zero offset, leaving the cell potential which, when the instrument is standardized, is directly converted into a pH, using the described linearity.

As long as hydronium ion concentrations are small, measurements with glass electrodes give good results. Otherwise, the measured concentration is usually less than the actual value, the two being related through the expression

$$a_x = \gamma_x C_x \qquad \textbf{(3-10)}$$

where a_x, mentioned previously, is the apparent concentration of substance X, C_x is its actual concentration, and γ_x is a parameter called the activity coefficient. When ISEs are used to detect low concentrations (their major use), the activity coefficient is approximately unity, making the actual and effective concentrations essentially equal.

When ISEs other than the hydrogen electrode are used, the usual procedure is to make up a series of standard solutions, encompassing the concentration of the analyte solution, and measuring the cell potentials for the standard and test solutions. A plot is then made of the cell potential (ordinate) versus log (C_x), where C_x is in any appropriate unit (usually molarity). If linear, the best straight line is drawn through all the points. If the plot is not linear, a curve is drawn through all the points. In either case, the analyte concentration in the test solution is determined from the calibration plot.

Safety Issues—Safety glasses must be worn at all times in a chemistry laboratory.

Procedure

1. **Collection of Rainwater** Collect a sample of rainwater (if possible), using a standard rain collector. Several weeks in advance of this experiment, preparations should be made for the collection, which is carried out when rain occurs. After collection the sample is stored in a refrigerator at 4°C until the experiment is carried out. At that time the sample must be allowed to come to room temperature.

2. **Calibration of the pH Meter** The pH meters will be set up and ready to be calibrated. When not in use, pH meters should be kept in the STANDBY mode. The electrode membrane (bulb) will be submerged in solution (often pH 7 buffer) and the cap on the side arm should have been removed. If not, do so.

   a. Calibrate the meter by using pH 7 and 10 buffers. Use DI water to rinse the electrode, catching the washings in a waste beaker. Blot the electrode dry and put into a beaker containing pH 7 buffer. Switch the meter to the MEASURE (or pH) position. If the meter does not read pH 7, adjust it to 7 by using the calibration dial. Gently swirl the buffer to help achieve equilibrium and a stable reading. (Refer to Notes section to see what must be done to obtain good results with a pH meter).

Switch the meter back into the STANDBY mode, rinse and dry the electrode as previously, and put the electrode into pH 10 buffer. Swirl gently and measure the pH. It should read pH 10; if it doesn't, adjust it to 10 using the TEMPERATURE COMPENSATION dial. To check the calibration, remeasure the pH 7 buffer. If it doesn't read pH 7, see the instructor.

b. To measure the pH of a sample, follow the previous procedure for rinsing and drying. Keep the meter in the STANDBY mode when the electrode is not in a solution. Gently swirl the solution under study to help achieve a stable reading on the meter.

c. Measure the pH of the rainwater sample, using the previous procedure for rinsing and drying the electrode. (If possible, a final rinsing of the electrode should be done with the next solution to be examined).

d. Take a sample of DI water that has been exposed to atmospheric carbon dioxide. DI water from a carboy or a squeeze bottle is a source. Measure the pH of the sample.

e. Measure the pH of six samples of widely varying values. These should include river water, well water, seawater (if available), bottled water, and three household samples (soda, wine, beer, Windex, coffee, tea, and milk are good samples).

f. Use 100 mL of one environmental sample to determine buffer capacity. Select a sample expected to be high in buffer capacity. Set up a clean buret and a magnetic stir plate. Fill the buret with 0.01000 M HCl (that will be provided) and take the initial volume reading. Add the sample to the titration beaker (250 mL) and measure the initial pH of the sample. Add 1 mL increments of 0.01000 M HCl. Measure the pH after each addition, stirring with a magnetic stir bar. Continue to a pH of about 4.0.

If the volume needed to reduce the pH by 1 unit is too small, increase the sample size and repeat the titration (called a potentiometric titration). Either 0.5 or 1.0 mL increments of titrant are added at a time. It is advantageous to plot pH (as ordinate) versus volume of titrant (as abscissa) as the titration progresses to determine where less (or greater) volume increments should be used.

Waste Minimization and Disposal

1. A variety of samples might be collected and studied in this experiment. Some samples may be considered hazardous waste. Your instructor will advise you about its disposal.

2. Fisher Scientific buffer solutions contain the following: pH 4 contains potassium acid phthalate; pH 7 contains monobasic hydrogen phosphate and sodium hydroxide; pH 10 contains potassium carbonate, potassium borate, and potassium hydroxide. The buffer solutions do not contain hazardous substances and can be disposed of in the laboratory drain (after checking with the lab instructor to see if this is acceptable).

3. Acids and bases can be flushed down the laboratory drain if their pH is between 3 and 11. If not, they can be diluted or neutralized first. (This also must be approved before executing.)

4. In making pH measurements use just enough solution to cover the bulb of the electrode to minimize the amounts of buffers and other solutions used.

Data Analysis

1. Prepare a plot of pH (ordinate) versus volume of titrant (a titration curve) for the environmental sample studied. (Do this with computer software, if possible.) Submit a copy of the titration curve.

2. Use the titration curve to determine the volume of titrant needed to decrease the initial pH by 1.00 unit. Use this volume to calculate the buffer capacity. The buffer capacity is the number of moles of the acid needed per liter of sample needed for this change in pH. Compare your result with the buffer capacity of a carbonate/bicarbonate system.

3. Tabulate your pH values, the type of sample, and expected values (if possible; some are available in the Introduction). Explain any differences between measured pH values and expected values. Relate the pH of the environmental samples to their location. Lake water will usually have a very different pH than a stream or the ocean.

Supplemental Activity

1. Do a second potentiometric titration for a second environmental sample, one expected to be significantly different than the first sample.

2. Include several types of bottled water samples in the pH measurements.

3. To test the student's work, a sample of known buffer capacity can be assigned and a potentiometric titration curve determined for the sample.

4. Water from a swimming pool has been suggested as an interesting sample for study in this, as well as the next, experiment.

5. Since rainwater and DI water are expected to be very low in ionic strength, it would be interesting to measure their pHs at various ionic strengths, including zero, 0.001, 0.010, and 0.100 M, using either NaCl or KCl.

Questions and Further Thoughts

1. Is buffer capacity the same if determined by titration with a strong base instead of a strong acid? Explain.

2. Can you develop a general equation, based on concentrations of acid and conjugate base, and pK_a for calculating buffer capacity? Show all work.

3. Is the amount of 0.0100 M HCl needed to decrease the pH of an environmental sample by two pH units two times that needed to decrease the pH by one unit?

4. Some water samples may be very low in buffer capacity. The authors have found tap water to be an interesting sample to study. A small amount of sediment or soil in the sample may increase the buffer capacity considerably.

5. The partial pressure of carbon dioxide in the pores of soil can be as great as 10 times that found in the atmosphere (due to plant and bacterial respiration). What would be the approximate pH of water in soil pores? How would this affect carbonate minerals in the soil?

Notes

Several factors affect the reproducibility and stability of pH readings. These are:

1. The pH meter must be well grounded. If a three-prong outlet is not available, use an adapter and ground the adapter wire to the screw in the center of the outlet plate.

2. The solution being tested must be mixed, either by frequent swirling or by using a magnetic stir bar and stirrer.

3. Many pH electrodes have a side arm for adding filling solution to the electrode. If your electrode is this type, the cap on the side arm of the electrode must be removed during measurements. The cap prevents evaporation of water from the filling solution in the electrode. The side arm must not be clogged with solid.

4. There must be sufficient filling solution within the electrode (at least one-half filled).

5. The solution being tested must completely cover the glass bulb of the electrode.

6. The electrode leads must be free of corrosion. If not, use fine sandpaper or a triangular file to remove the corrosion.

7. Read the specific instructions given with your pH meter to account for test sample temperatures being different from 25°C.

8. Read the instructions given with your pH meter to calibrate the meter. The use of two buffers for calibration is recommended.

9. Whenever a pH electrode is removed from solution the meter must be in STANDBY mode.

Literature Cited

1. C. Baird, *Environmental Chemistry*, W. H. Freeman, New York, 1995.

2. T. L. Brown, H. E. LeMay, Jr., and B. E. Bursten, *Chemistry, the Central Science,* 7th ed., Prentice-Hall, Upper Saddle River, NJ, 1997.

3. T. E. Larsen and L. M. Henley, Determination of low alkalinity or acidity in water, *Anal. Chem.*, *27(5)*:851–852(1955).

4. S. E. Manahan, *Environmental Chemistry*, 5th Edition, Lewis Publishers, Chelsea, MI, 1991.

5. G. H. Schenk, R. B. Hahn and A. V. Hartkopf, *Introduction to Analytical Chemistry*, 2nd ed., Allyn and Bacon, Boston, 1981.

6. T. G. Spiro and W. M. Stigliani, *Chemistry of the Environment,* Prentice-Hall, Upper Saddle River, NJ, 1996.

Experiment 4

Alkalinity of Streams and Lakes

Objective—The objective of this experiment is to introduce a fundamental and very important property of natural waters, the alkalinity. There is a significant difference between alkalinity and basicity, which will be discussed.

Introduction— For several reasons it is important to determine the capacity of a body of water to neutralize acid (hydronium ion). A lake under the influence of acid rain or mine water runoff can absorb only so much acid before its pH begins to drop significantly. The capacity of a water sample to absorb hydronium ion is its alkalinity. Basic water is water that has a pH greater then 7. Alkalinity is a type of buffering capacity and thus a capacity (extensive) factor, whereas basicity is an intensive factor, independent of the quantity of sample.

The pH of many natural waters, as we saw in Experiment 3, is close to 8, and changes in pH are resisted by the presence of a carbonate/carbon dioxide buffer system. Carbon dioxide in water undergoes the following complex equilibria:

$$CO_2\,(g) + H_2O(l) \rightleftharpoons CO_2\,(aq)$$

$$CO_2\,(aq) + H_2O\,(l) \rightleftharpoons H_2CO_3\,(aq)$$

$$H_2CO_3\,(aq) \rightleftharpoons H^+\,(aq) + HCO_3^-\,(aq)$$

$$HCO_3^-\,(aq) \rightleftharpoons H^+(aq) + CO_3^{2-}\,(aq)$$

Frequently present in the geochemical environment of various water systems are metal carbonates, especially calcium carbonate. In water, calcium carbonate is only slightly soluble due to its small K_{sp} (4.6 x 10^{-9} at 25°C). However, due to the hydrolysis of carbonate ion,

$$CO_3^{2-}\,(aq) + H_2O\,(l) \rightleftharpoons HCO_3^-\,(aq) + OH^-\,(aq)$$

there is an increase in the solubility of calcium carbonate and, therefore, an increase in pH.

If carbon dioxide is also present in the body of water, the following reaction can also occur,

$$H_2O\,(l) + CO_2\,(aq) + CO_3^{2-} \rightleftharpoons 2\,HCO_3^-\,(aq)$$

and enhances the solubility of both carbon dioxide and calcium carbonate (Le Châtelier's principle).

When carbon dioxide and calcium carbonate are both present, if a strong acid is introduced into the system, the following reaction occurs,

$$H_3O^+(aq) + HCO_3^-\,(aq) \longrightarrow 2H_2O(l) + CO_2\,(g)$$

If strong base is added,

$$OH^-\,(aq) + HCO_3^-\,(aq) \longrightarrow H_2O\,(l) + CO_3^{2-}(aq)$$

Thus, such systems act as buffers where HCO_3^- is the acid and CO_3^{2-} is the conjugate base.

High alkalinity water contains elevated levels of dissolved solids. Although this may be undesirable for boiler water and food processing, it serves as a source of inorganic carbon for the growth of algae and other aquatic life. Hence, alkalinity is a measure of the fertility of the water.

Algae extract the carbon dioxide they need for photosynthesis from aqueous carbon dioxide,

$$xCO_2\,(aq) + x\,H_2O(l) + h\nu \longrightarrow (CH_2O)_x + xO_2(aq)$$

In the absence of sufficient carbon dioxide, bicarbonate dissociates,

$$HCO_3^-\,(aq) \rightleftharpoons CO_2(aq) + OH^-\,(aq)$$

and as algae use bicarbonate, the body of water becomes more basic.

Theory—Alkalinity can be defined as the moles of monoprotic acid per liter needed to react completely with all base present. Since alkalinity is generally due to carbonate, bicarbonate and hydroxide, it is defined by

$$[\text{alkalinity}] = 2[CO_3^{2-}] + [HCO_3^-] + [OH^-] - [H_3O^+]$$

where the brackets refer to molar concentrations and the coefficients result from the number of H^+ ions the base can react with (for example, CO_3^{2-} reacts with $2H^+$). Usually the hydronium ion concentration is small enough to be neglected if the water has a basic pH. In fact, $[OH^-]$ is usually small compared with the first two terms and can be neglected. There are a few minor contributors to alkalinity, including ammonia and the conjugate bases of phosphoric, boric, silicic, and organic acids.

The alkalinity can be expressed in units of mg/L of calcium carbonate, based on the reaction,

$$CaCO_3 + 2\,H^+ \longrightarrow Ca^{2+} + CO_2 + H_2O$$

If the water is very basic (pH greater than 8.3), a quantity called the phenolphthalein alkalinity can be determined. This is mainly due to the presence of the carbonate ion. In this case the sample is titrated with standard acid to a pH of about 8.3 (where phenolphthalein goes from pink to colorless). Most environmental waters have a pH less than 8.3 and thus have a zero phenolphthalein alkalinity.

When the pH of a sample is less than 8.3, methyl orange is used as the indicator and the water sample is titrated with acid to a pH of 4.3. At this point all the HCO_3^- has been converted to CO_2. Figure 4-1 shows how the pH varies with titrant volume when a solution of Na_2CO_3 is titrated with acid.

Figure 4-1 Titration of 25 mmol of Sodium Carbonate with HCl

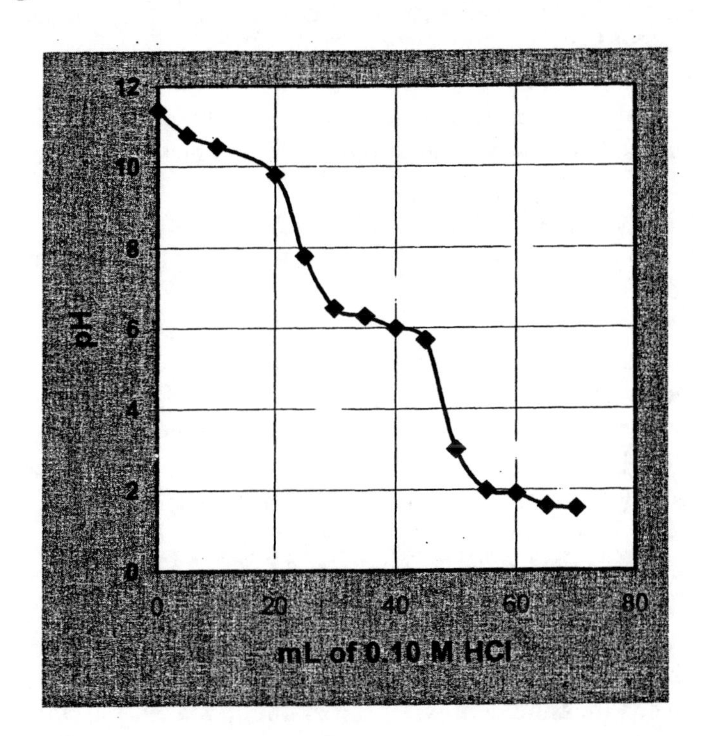

Typical seawater carbonate alkalinities are about 1.2 mmol/L. Anaerobic digesters (used in treating sewage sludge) typically have supernatant alkalinities of 2000–4000 mg/L $CaCO_3$.

Safety Issues

1. Safety glasses must be worn at all times in the chemistry laboratory.

2. Avoid skin and eye contact with sodium hydroxide and hydrochloric acid. If contact is made, rinse your

Procedure—The sample should not be filtered, diluted, or concentrated before testing. Since a large range of alkalinity is possible, a rough measurement is made initially using 0.1 M HCl titrant. This allows for adjusting sample size so that the titration volume is greater than 10 mL, but less than 50 mL. If for this sample the titrant volume used is very small, a 0.02 M HCl solution is used with an appropriate sample size. Thus, the "test titration" with 0.1 M HCl is a guess to see what concentration of acid is needed.

Use the following procedure:

A: Standardization

1. Prepare a 0.1 M sodium hydroxide solution. Quickly weigh about 4 g of sodium hydroxide pellets using a small beaker and transfer to a 1 L plastic bottle. Fill the bottle with 1 L of DI water and mix thoroughly. This solution will be standardized. (An alternative procedure is to dilute 8 g (about 5–6 mL) of 50% (w/w) sodium hydroxide to 1 L with DI water. This procedure eliminates sodium carbonate as an impurity since it is insoluble in the concentrated base.)

2. Prepare 0.1 M HCl (for alkalinity greater than 20 mg/L). In a fume hood, measure out 8.3 mL of concentrated HCl using a 10 mL graduated cylinder and dilute to 1 L in a glass or plastic bottle. Mix well.

3. Prepare 0.02 M HCl. Dilute 200 mL of the 0.1 M HCl to 1 L using volumetric flasks. Standardization of this solution may be done, but is not necessary. The HCl can be measured out with a graduated cylinder that can be calibrated, if necessary.

4. Standardize the 0.1 M NaOH against primary standard potassium acid phthalate (KHP). Weigh accurately (to 0.1 mg) three samples of KHP (previously dried) weighing about 0.5 g each (0.49–0.51 g). Quantitatively transfer the KHP to 250 mL Erlenmeyer flasks and dissolve in about 75 mL DI water. Add 3 drops of phenolphthalein indicator and titrate with the 0.1 M NaOH until the faintest pink persists for 30 seconds. The buret must be rinsed with three 10 mL portions of the sodium hydroxide before use.

5. Use the standardized 0.1 M NaOH solution to titrate 25.00 mL (pipetted) aliquots of the 0.1 M HCl diluted to about 75 mL with DI water. Do three determinations of the molarity of the HCl and use the average value in subsequent calculations. (An exercise at this point to get a feel for the number of trials needed to obtain a reasonable standard deviation is to continue titrating 25.00 mL aliquots and calculating the standard deviation after each measurement [after three] until a reasonable standard deviation is obtained. That is, if the average molarity is 0.102 then the standard deviation might be 0.003 M, and the relative standard deviation is about 3%.)

B: Indicator Titration for Alkalinity

1. Pipet 100 mL of sample (or the appropriate amount determined in a test titration) into a 250 mL Erlenmeyer flask. Add 3–5 drops of methyl orange.

2. Rinse a buret with three 10 mL portions of 0.1 M HCl. Fill with the acid and record the initial volume. Titrate the sample with the standardized 0.1 M HCl to the end point (which is orange to red), and record the final volume.

3. If the alkalinity is less than 20 mg/L, as determined by the test titration, use 0.02 M HCl and adjust the sample size if necessary. Titrate the sample as in Parts 1-2.

4. Do two additional titrations on the same sample according to Steps 1 and 2.

5. Repeat Steps 1–3 for a second type of environmental sample.

C: Potentiometric Titration

1. Calibrate the pH meter according to the procedure of Experiment 3.

2. Choose one of the environmental samples studied by the indicator titration to examine in this part of the experiment. Use a pipet to measure out the appropriate amount of sample into a 250 mL beaker. Lower the pH electrode into the sample, being certain that the bulb of the glass electrode is completely covered.

3. Using a magnetic stir bar and plate, obtain a potentiometric titration curve by adding standard 0.1 M HCl from a buret in either 0.5 mL or 1.0 mL increments, stirring and measuring the pH, until a pH of 4.0 is obtained. Record the pH after each addition of titrant.

Waste Minimization and Disposal

1. The water sample used for determining buffer capacity need not be collected from a polluted body of water and thus need not be hazardous waste. Otherwise, your instructor will direct you on the disposal method.

2. Extra hydrochloric acid and sodium hydroxide that needs to be disposed of can be used to neutralize one another. A final pH between 3 and 11 is needed before disposal in the laboratory drain. Standard acids and bases, however, might be used in later experiments and should not be disposed of before checking with the instructor.

Data Analysis

1. The molarity of sodium hydroxide is obtained from the stoichiometric titration reaction, NaOH + KHP $\longrightarrow$ H_2O + KNaP. At the end point the moles of base are equal to the moles of acid. Also, mol NaOH = [(mL)/1000](M) and mol KHP = mass/molar mass. Therefore, the molarity of NaOH is given by,

$$\mathbf{M = (mass\ KHP / 204.23) / (mL\ NaOH / 1000)}$$

Report each individual molarity and the average of the values for NaOH.

2. Report the titration volumes and the calculated molarities of the 0.1 M HCl, and report the mean molarity.

2. Report the titration volumes and the calculated molarities of the 0.1 M HCl, and report the mean molarity.

3. If the 0.02 M HCl was used, report its molarity (from standardization or by calculation from the dilution of a standard solution).

4. The alkalinity for both the indicator and potentiometric methods is given by

$$\textbf{[Alk]} = \textbf{½(mL HCl)(}\mathbf{M_{HCl}}\textbf{) (100.0 mg / mmol) / L of sample}$$

when the alkalinity is expressed in mg $CaCO_3$/L. The volume of titrant is the volume of HCl needed to achieve a pH of 4.3 and 100.0 mg/mmol is the molar mass of $CaCO_3$ (g/mol = mg/mmol).

a) Report the alkalinity of the water sample that was done with one replicate. Report the individual values and the mean value. Compare with the potentiometric method done on the same sample.

b) Report the alkalinity for the second sample type done using the indicator method.

5. Compare the alkalinities for the two sample types and discuss their difference in terms of the sample site location.

6. If Experiment 3 has been done, use the results for the potentiometric titration curve to obtain another value of the alkalinity, and include this value in the previous discussions.

Supplemental Activity

1. Total dissolved solids were determined in Experiment 1, and if the same samples are used in this experiment, it may be possible to determine if there is a relationship between dissolved solids and alkalinity. (In Experiment 8 the water hardness, which is partly due to calcium ions, will be determined. If the hardness is done using the same samples as in this experiment, it may be possible to relate the hardness to the alkalinity.)

2. Bottled spring water samples would make an interesting addition and comparison for this experiment.

3. It would be interesting to determine the pH of the samples studied in this experiment and list these values in a table with the determined alkalinities.

Questions and Further Thoughts

1. What is the difference between alkalinity and basicity?

2. Can an acidic solution have a measurable alkalinity? Explain with an example.

3. What is the difference between alkalinity and buffer capacity?

Notes

1. Sodium carbonate can be used to standardize the hydrochloric acid instead of using standard base.

2. A CBL (calculator-based laboratory) unit can be used to store pH data, which can then be downloaded to a computer to generate the potentiometric titration curve. Details about this procedure are given in the instructor's manual.

3. The density of 50% sodium hydroxide solution is 1.52 g/mL at room temperature.

Literature Cited

1. Standard Method 310.1, *Methods for Chemical Analysis of Water and Wastes,* EPA-600/4-79-020, Office of Research and Development, U.S. Environmental Protection Agency, Cincinnati, OH, 1979.

2. C. Baird, *Environmental Chemistry,* Freeman, New York, 1995.

3. T. L. Brown, H. E. LeMay, Jr., and B. E. Bursten, *Chemistry, The Central Science,* 7th ed., Prentice-Hall, Upper Saddle River, NJ, 1997.

4. A. E. Greenberg, L. S. Clesceri and A. D. Eaton, Eds., *Standard Methods for the Examination of Water and Wastewater,* 18th ed., American Public Health Association, Washington, DC, 1992.

5. T. E. Larsen and L. M. Henley, Determination of low alkalinity or acidity in water, *Anal. Chem.*, 27(5):851–852 (1955).

6. S. E. Manahan, *Environmental Chemistry,* 5th ed., Lewis Publishers, Chelsea, MI, 1991.

7. G. H. Schenk, R. B. Hahn and A. V. Hartkopf, *Introduction to Analytical Chemistry,* 2nd ed., Allyn and Bacon, Boston, 1981.

8. T. G. Spiro and W. M. Stigliani, *Chemistry of the Environment,* Prentice Hall, Upper Saddle River, NJ, 1996.

9. J. F. J. Thomas and J. J. Lynch, Determination of carbonate alkalinity in natural waters, *J. Am. Water Works Assn.*, 52:259 (1960).

Experiment 5

Using Ion-Selective Electrodes to Determine Trace Levels of Ions in Natural Waters

Objective—The objective of this experiment is to learn the use of and carry out applications of ion-selective electrodes (ISEs) in measuring part per million levels of the simple monovalent ions potassium (K^+) and fluoride (F^-).

Introduction—Most cations and anions found in natural waters are not present in large amounts. Tables 5-1 and 5-2 show typical ion concentrations in seawater and North American streams, respectively. Some ions, including Ca^{2+}, Mg^{2+}, and Cl^- are plentiful enough to be analyzed for by "wet" methods of analysis; that is, by titration or gravimetry. Although sodium ion is plentiful, there is no simple wet analytical method for its analysis. Sodium ion does not complex easily, nor does it undergo any other simple reaction which might be made the basis of a wet analytical procedure. Potassium and other alkali metal ions are similar in this respect, but are much less abundant. Flame absorption and emission are often the usual analytical methods for the alkali metals.

Halide ions other than chloride are present at very low levels in environmental waters and thus cannot be easily measured. Fluoride, found at higher levels in dental products, can be determined using a colored complex as the basis of a colorimetric method.

Ion-selective electrodes, introduced in Experiment 3, can be used to detect very low ion concentrations. For monovalent cations a glass electrode, similar to the pH electrode, can be used. The particular ion to which the electrode responds is determined by the composition of the glass. Generally, such electrodes can detect only Li^+, Na^+, K^+, Rb^+, Cs^+ and NH_4^+. Anions, including F^-, S^{2-}, Cl^-, Br^-, and I^-, are measured with solid-state electrodes where the membrane has a composition that includes the ion to be measured. A membrane of PbS, for example, can be used to detect either lead(II) or sulfide ions. The advantages of such electrodes are their great specificity (at least towards anions—cations are not quite as free from interferences), their sensitivity at very low concentrations, their portability, and the speed of analysis.

The potassium ion, which we study in this experiment, illustrates another use of the glass electrode (in addition to pH measurement), and the analysis of environmental fluoride illustrates the use of a solid-state electrode. Both potassium and fluoride ions are important constituents of natural waters as shown in Tables 5-1 and 5-2.

A: Potassium in the Environment — Potassium is the seventh most abundant element and makes up 2.6% of the earth's crust. In seawater it is present at 400 ppm, but in potable water it seldom reaches 20 ppm. The weathering of potassium feldspar, a primary mineral, is one of the most important sources of potassium ion in natural waters. It is also responsible for the formation of clay in soils as shown in the equation,

$$3KAlSi_3O_8 + 2CO_2(aq) + 14H_2O \longrightarrow 2K^+(aq) + 2HCO_3^-(aq) + 6H_4SiO_4 + KAl_3Si_3O_{10}(OH)_2$$

feldspar

Table 5-1 The Concentrations of the Most Abundant Ions in Seawater

Species	M(mol/L)	g/kg	Mg/kg
Cl^-	0.55	19.35	------
Na^+	0.47	10.76	------
SO_4^{2-}	0.028	2.71	2710
Mg^{2+}	0.054	1.29	1290
Ca^{2+}	0.010	0.412	412
K^+	0.010	0.40	400
$CO_2(aq)$	2.3×10^{-3}	0.106	106
Br^-	8.3×10^{-4}	0.067	67
Sr^{2+}	9.1×10^{-5}	0.0079	8
F^-	7×10^{-5}	0.001	1

Source: T. L. Brown, H. E. LeMay, Jr., and B. E. Bursten, *Chemistry, the Central Science*, 7th ed., Prentice-Hall, NewYork, 1997.

Table 5-2 The Concentrations of the Most Important Ions in River Water in the United States

Ion	M(μmol/L)	mg/kg
HCO_3^-	960	60
Ca^{2+}	380	15
Mg^{2+}	340	8
Na^+	270	6
Cl^-	220	8
SO_4^{2-}	120	12
K^+	59	2
F^-	5	0.1

Source: C. Baird.Environmental Chemistry, Freeman, New York, 1995.

Potassium is an essential plant nutrient, and relatively high levels are absorbed by growing plants. It activates certain enzymes and plays a role in the water balance in plants. Crop yields are lower in potassium-deficient soils, and potassium may become a limiting nutrient in soils heavily fertilized with other nutrients. In streams impacted by heavy fertilizer usage on adjacent farmlands, potassium concentrations are anticipated to be higher than usual background levels.

B: Fluoride in the Environment—Like many elements, including copper and zinc, fluoride is essential to plants and animals at trace levels, but is deleterious (and even lethal) at elevated levels. Fluoride levels in the environment are highly variable. It occurs widely distributed in the earth's crust. The source of most fluoride is the weathering of fluorapatite, $Ca_5F(PO_4)_3$. However, it is found in many natural substances, including coal, clay, and various minerals. The chief fluoride-containing minerals are fluorspar, cryolite (Na_3AlF_6), and fluorapatite. Fluorspar (CaF_2) is the chief mineral on which the production of fluorine is based, due to its abundance. Deposits of fluorspar are spread throughout Illinois and Kentucky. Another significant source of fluoride results from the processing of phosphate rock for fertilizers.

Anthropogenic sources of fluoride in soils are chiefly fertilizers, especially superphosphates. Fluorides are also used as pesticides, mainly to control ants, roaches, and rodents. Due to the persistence of fluorides, pesticides and fertilizers may contribute significantly to the fluoride level in soils. Fluorides have been used as disinfectants in breweries and distilleries and also as a wood preservative.

In communities where fluoride levels in drinking water are low, a soluble fluoride salt (usually sodium fluoride) may be added to bring the level to about 1 ppm (approximately 50 μM). This concentration has been found to be optimal in preventing tooth decay.

Fluorides in Humans and Other Animals—Fluoride uptake by humans occurs mainly from food and water, whereas inhaled amounts are quite small. It accumulates primarily in teeth, bone, and the soft tissues where it is thought to replace the hydroxy group (—OH) of hydroxyapatite, a mineral phase occurring during the formation of bone and tooth enamel. The resulting material is fluorapatite, which is harder and more inert than hydroxyapatite, and thus is important in preventing tooth decay.

Once absorbed, a part of the fluoride deposits in the skeletal tissue, and a part is excreted in the urine. Skeletal sequestration (a sink) and renal excretion are the two major pathways by which the body prevents the accumulation of toxic levels of fluoride.

Fluoride is transported in human blood in both inorganic and organic forms. Inorganic fluoride is an enzyme inhibitor that acts by forming metal-fluoride-phosphate complexes. These interfere with the activity of enzymes that require a metal ion cofactor. Fluoride inhibits the energy-production system of the cell.

At elevated levels, fluoride is toxic, and possibly carcinogenic. At about the 10 ppm level, it results in mottled teeth and at yet higher concentrations causes bone anomalies. Ingesting soluble fluoride results in the formation of hydrogen fluoride (due to gastric HCl), causing irritation of the gastric mucosa. Fluorosis, a condition affecting bones, kidney function, and nerve and muscle function, results from the ingestion of elevated levels of fluoride.

Fluoride has been shown to accumulate in animals that consume fluoride-containing foliage, but there is no evidence of biomagnification in the food chain.

Fluorides in Food—Foods grown in soils with high background levels of fluoride generally have increased levels of fluoride. The fluoride level in foods is affected by the use of fertilizers and fluoride-containing pesticides. Some plants, especially tea, accumulate fluoride, concentrating it in their leaves and stems. Other foods having elevated fluoride levels include all types of seafood (2–29 mg/kg) and bone products (such as bone meal and gelatin).

In areas where fluoride is not added to drinking water, the total intake from food and water is usually less than 1 mg/day. Where fluoridation is practiced, the average intake is about 2.7 mg/day. The ingested fluoride from toothpastes and other fluoridated dental products can amount to 0.25 mg/day.

Fluorides in Soils—For soil and water, the most important natural source of fluoride is the weathering of fluoride-containing rocks. Other sources of fluoride in soil are the deposition of atmospheric fluoride and plant and animal wastes.

The fluoride level in soils ranges generally from 200 to 300 ppm. However, in soils where fluoride-containing minerals are abundant, the fluoride level may be much higher. In the United States, the average level in soils is 340 ppm on the east coast and 410 ppm on the west coast. Fluoride levels tend to increase with soil depth, with typical results of 190 ppm for depths from 0 to 8 cm and 292 ppm for depths between 7 and 30 cm. Soluble fluorides are absorbed by vegetation; both anhydrous and aquatic forms of HF are readily absorbed by plants.

The expected forms of fluoride at hazardous waste sites include NaF, CaF_2, and HF. At low pH, fluoride forms HF (pK_a = 3.13) and at pH > 10, Al^{3+} forms strong complexes with fluoride. Soluble fluorides are highly adsorbed on soils as CaF_2 at pH 6.5 or greater if sufficient $CaCO_3$ is present.

Fluoride in Water—The level of naturally occurring fluoride in surface waters is 10–300 ppb. Higher levels are found in groundwater, with a range of 20–1500 ppb. Waters with high salinities, such as estuaries, have elevated fluoride levels. The Great Salt Lake has a fluoride concentration of 14 ppm.

Seawater has about 10 times the fluoride level found in freshwater. The precipitation of $CaCO_3$ dominates the removal of F^- from seawater. The average fluoride concentration in seawater is about 1 ppm, whereas in U.S. rivers it is about 100 ppb. In the authors' local environmental waters the fluoride level is about 500 ppb, about the same level as found in the flesh of fish from the same waters.

Fluoride in the Atmosphere—The largest natural source of fluoride in the atmosphere is volcanic eruptions, and HF is the primary form of fluoride. The weathering of rocks and minerals containing fluoride contributes to the atmospheric load of fluoride. The marine aerosol is also an important source of tropospheric HF. Stratospheric HF can form by the photolytic decomposition of chlorofluorocarbons (CFCs or Freons).

Theory The Potassium Glass Electrode—For the potassium electrode the Nernst equation (Eq. 3-6) becomes,

$$E_{K^+} = E^\circ_{K^+} + 0.05915 \log(a_{K^+}) \qquad (5\text{-}1)$$

at 25°C. Since the concentration of the potassium ion is usually small, its activity can be replaced by its molar concentration, $[K^+]$. The constant $E^\circ_{K^+}$ is the standard electrode potential ($a_{K^+} = 1$) at 25°C for the reduction half-reaction,

$$K^+ (aq) + e^- \longrightarrow K(s) \qquad E^o_{K^+} = -2.925 \text{ V}$$

The standard electrode potential is an important parameter that gives a quantitative indication of the driving force (or free energy decrease) for the reduction half reaction to occur. It is a function of temperature.

The potassium glass electrode is similar to the pH electrode, and it may have an internal reference electrode, as in a pH combination electrode, or it may be used in conjunction with an external reference electrode, such as a calomel electrode. In a glass electrode selectivity results from the presence of anionic sites on the glass surface that show an affinity for certain cations. By varying the relative amounts of Na_2O, Al_2O_3, B_2O_3, and SiO_2, it is possible to produce glasses that respond mainly to one ion, usually an alkali metal ion, ammonium, or silver ion.

A potassium glass electrode is calibrated by measuring the potentials of a series of solutions of known potassium ion concentration. The usual range is 0.1–1000 ppm, but higher concentrations can be measured by first diluting the sample to the appropriate range. In carrying out an analysis it is important that the ionic strength of the standards and samples be kept nearly constant. The ionic strength is a quantitative measure of the sample matrix (if the analyte background is chiefly ionic in nature). The ionic strength is defined as $\mu = \frac{1}{2}\Sigma C_i Z_i^2$, where Z_i is the charge on ion i and C_i is its activity (or molarity). An ionic strength adjustor (ISA) is added to both the samples and standards to give a constant electrostatic background matrix that negates the effects on the potential due to the presence of other ionic solutes in the sample. The ISA for a potassium ISE is 5 M sodium chloride (see Note 6).

Electrode potentials are measured using either a pH meter with an expanded millivolt scale or a specific ion meter having a direct concentration scale.

The Fluoride Ion Solid-State Electrode. Negative ions, including F^-, Cl^-, Br^-, I^-, and S^{2-}, can be determined using solid-state electrodes, where the membrane has a composition which includes the ion to be determined. Solid-state membranes are selective toward anions in the same way that some glasses are selective to specific cations. A glass owes its specificity to the presence of anionic sites on its surface, and a solid-state membrane having similar cationic sites might be expected to respond selectively toward anions. These membranes are insoluble salts of the anion of interest and a cation that selectively precipitates that anion from aqueous solution. Thus silver halides are used to detect various halide ions.

In addition to strength and resistance to corrosion, solid-state electrodes must have sufficient electrical conductivity to allow a potential measurement to be made. The fluoride electrode consists of a single crystal of lanthanum fluoride doped with a rare earth metal to increase conductivity. The membrane separates an internal reference solution at the inner surface and the solution to be measured at the outer surface. Alternatively, direct contact from the reference lead can be made to the membrane. The electrode can be used with an external calomel reference electrode. A typical fluoride electrode is illustrated in Figure 5-1.

Figure 5-1 Solid State Fluoride Ion-Selective Electrode

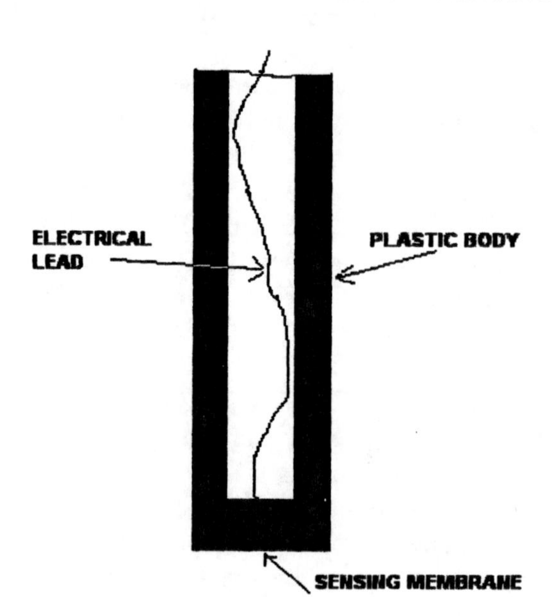

The fluoride electrode shows the theoretical response to changes in fluoride ion activity; that is, it obeys the Nernst equation in the form,

$$E_{F^-} = E°_{F^-} - 0.0591 \log(a_{F^-}) \qquad (5\text{-}2)$$

This equation is valid to an activity as low as 1 μM in F^- and is selective to fluoride over other common anions by several orders of magnitude. Only hydroxide interferes significantly, but this can be avoided by using a buffer to decrease the hydroxide concentration to a low value.

Another limitation of the fluoride electrode is that it can only detect the fluoride ion itself and not any of its combined forms, one of which is HF. However, it is not difficult to free F^- from sample matrices. Interferences result from cations that can complex F^-, including Ca^{2+}, Mg^{2+}, Sr^{2+}, and others. Adding an appropriate buffer not only controls pH and converts HF into F^-, but also releases F^- from its metal complexes. The buffer used in fluoride ion measurements by ISE thus accomplishes the following:

a. It provides a given, uniform ionic strength.

b. It adjusts the pH to >5, freeing F^-, preventing the loss of HF.

c. It prevents interference due to hydroxide by keeping the pH low.

d. It breaks up metal-fluoride complexes (including Al–F and Fe–F complexes) into F^-.

Analysis with the fluoride electrode is applicable to less than 0.1 ppm to more than 10 ppm.

Safety Precautions

1. Safety glasses must be worn at all times in the chemistry laboratory.

2. Sodium fluoride is very toxic and possibly carcinogenic. Exercise care in handling the solid and its solutions and avoid contact with bare skin.

3. Exercise great care in handling solid sodium hydroxide and its solutions. Avoid all skin contact, but if it occurs, rinse the exposed areas for several minutes in cold water.

Procedure

A: Potassium Ion For convenience, aqueous samples are studied in this experiment. The runoff from an agricultural area would provide a good study sample. Samples of seawater, brackish water, well water, fruit juices, and bottled waters provide interesting studies.

Sample Preparation - Samples and standards should not be stored in soft (flint) glass containers due to leaching of K^+ from the glass. If possible, collect samples from the environment in plastic containers that have been thoroughly cleaned. Use the following procedure to measure K^+ in aqueous samples.

1. Prepare 100 mL each of a series of standard KCl solutions containing 0.1000, 1.000, 10.00, and 100.0 ppm K^+ by making appropriate dilutions of a 100.0 ppm (or a 1000 ppm) stock solution (which may be provided).

2. To 50 mL of 0.1000 ppm standard KCl in a 150 mL beaker, add 1 mL of the ISA and enough 6 M NaOH (added dropwise) to bring the pH to about 11 (measured with pH paper). Hydronium ion is a serious interference of the potassium electrode and a high pH avoids the problem. Stir the solution, immerse the electrode(s), swirl or stir for 2 min to stabilize the potential. Finally, record the potential to 0.1 mV.

3. Rinse the electrode(s) with DI water, blot dry and repeat the previous measurement using the next most concentrated standard. Continue with the rest of the standards.

4. Repeat the previous procedure for at least five environmental samples.

B: Fluoride Ion Other than treating with buffer, no other treatment of aqueous samples is necessary.

Collect five samples for testing. The same samples used for potassium can be used in this part of the experiment. Samples of varying salinity and hardness would be of interest as well as water from streams and ponds where there may be runoff from agricultural areas. Fluoride is also elevated in areas where there have been forest fires and wood preservation. Also test a sample of drinking water that has been fluoridated and a sample of water that has not been treated with fluoride.

Use the following procedure to test aqueous samples for fluoride:

1. Dissolve 0.221 g of NaF in 1 L of DI water to obtain a stock solution that is 100.0 ppm in F^-. This stock solution may be provided.

2. Prepare a series of standard fluoride solutions by pipetting 0.50, 1.00, 2.00, and 4.00 mL of the 100 ppm stock solution into each of four 100 mL volumetric flasks and diluting with DI water to 100 mL. This yields standard solutions with 0.50, 1.00, 2.00, and 4.00 ppm fluoride.

3. Pipet 25 mL of the most dilute standard into a 150 mL beaker and add an equal volume of buffer. (See Note 5 for the buffer composition.) The total volume must be sufficient to immerse the electrodes and at the same time allow the magnetic stir bar to freely rotate.

4. Measure the potential, using the expanded mV scale, while gently stirring the solution. Allow the electrode(s) to remain in the solution until a constant reading is obtained.

5. Remove the electrode(s), rinse with DI water, blot dry, and repeat Steps 3 and 4 for the next standard solution.

6. Continue Steps 3–5 for all standards and samples, and then repeat the measurements for one standard and one test sample to check the reproducibility of the measurements.

Waste Minimization and Disposal

1. Only about 100 mL of water sample is needed for analysis by ion-selective electrode. However, if samples are going to be used for subsequent experiments, 0.5 L should be collected. Extra water can be disposed in the laboratory drain if it is not hazardous waste. If it is, follow your instructor's directions on its disposal.

2. The potassium chloride standards are not hazardous and can be rinsed down the laboratory drain, if necessary.

3. Although fluoride is hazardous at high levels, it is usually present in the environment at the levels of the standard solutions used in this experiment. However, do not rinse them down the laboratory drain unless your instructor so indicates.

Data Analysis

A: Potassium Ion

1. Prepare a table with all values of measured potentials and sample types.

2. Prepare a plot of E_{meas} as ordinate versus log (ppm K^+) as abscissa. Use a linear least-squares analysis to

calculate the slope and y-intercept of the plot. Compare with the theoretical value after converting Equation 5-1 into an equation where activity has been converted into ppm units. Also, obtain the correlation coefficient and comment on the linearity.

3. Use the previous plot (with interpolation) or computer software to calculate the potassium ion concentration for each sample. Tabulate the values. (Alternatively, your instructor may want you to calculate the equation of the straight line from the least-squares analysis and use the equation to calculate concentrations from potentials. This is more accurate than using a plot directly).

4. Discuss the results for the concentrations of potassium ion as related to the sample type, location and other environmental parameters, such as salinity.

B: Fluoride

Repeat the previous data analysis for K^+ using fluoride instead. Discuss the results for drinking water and indicate whether your results are reasonable. Compare the least squares slope with the theoretical slope of Eq. 5-2.

Supplemental Activity

1. A fluoride analysis of toothpaste can be carried out without much additional work. A 1 g sample of toothpaste with about 0.5% stannous fluoride is dissolved in DI water and diluted to 100 mL. The resulting solution can be measured directly with the fluoride electrode without buffer, and then again with the buffer. This tests the need for the buffer solution for this particular analysis. The pH of the toothpaste solution should also be determined.

2. Bottled water samples make an interesting study for both the potassium and fluoride ions.

Questions and Further Thoughts—Fluoride is not easy to extract from plant material. The sampling of plant leaves for fluoride may be of interest since HF accumulates in leaves. However, heating such samples with acid (for digestion) results in a loss of HF. Ashing a sample in a covered crucible with a Meker burner can be followed by treatment with acid. The HF is recovered by heating the digestate/acid mixture in a distillation apparatus and capturing the volatile acid in sodium hydroxide, somewhat like a Kjeldahl determination of nitrogen in plants.

Notes

1. A lead ISE electrode can be used in place of the fluoride electrode to illustrate a solid-state electrode. Lead is plentiful in sediments or soils near roadways, and decreases with increasing distance from the road. In this case an acid digestion with very high-purity nitric acid is carried out, followed by cooling, filtering, and diluting to volume in a volumetric flask. Generally a 5 g sample will provide a measurable concentration of lead when diluted to a volume of 100 mL.

2. The authors have found the sodium glass electrode to be stable yielding results that are interesting to discuss, including studies of salt-water intrusion.

3. Some bottled waters have considerable levels of sodium.

4. When analyzing for fluoride ion using an expanded-scale pH meter or selective-ion meter, frequently recalibrate the electrode by checking the potential reading of the 1 ppm fluoride standard and adjusting the calibration control, if needed, until the meter reads as previously.

5. In addition to adjusting the pH to free fluoride from HF and to eliminate any interference due to OH^-, the buffer also provides a constant background ionic strength and breaks up fluoride complexes. The buffer can be prepared using the following procedure.

To 500 mL DI water in a 1 L Erlenmeyer flask add 84 mL of concentrated HCl, 242 g TRIS (hydroxymethyl) amino methane, and 230 g sodium tartrate. Use a magnetic stir bar and plate to stir and dissolve and allow to cool to room temperature. Dilute to the 1 L mark with DI water and transfer to a bottle.

6. The ISA for the potassium ISE is 5 M NaCl. This is made by dissolving 29.2 g of reagent grade NaCl in 50 mL DI water, followed by dilution to 100 mL in a 100 mL volumetric flask.

Literature Cited

1. C. Baird, *Environmental Chemistry,* Freeman, New York, 1995.

2. M. E. Bell, E. J. Largent, and T. G. Ludwig, *Fluorides and Human Health,* World Health Organization Monograph Series, 17-29, Geneva, 1970.

3. H. J. M. Bowen, *Trace Elements in Biochemistry,* Academic Press, New York, 1966.

4. H. A. Cook, Fluoride in tea, *Lancet 2*:329 (1969).

5. B. Eyde, Determination of fluoride in plant material with an ion-selective electrode, *Fresenius Z. Anal. Chem.*, 311:19-22 (1982).

6. R. B. Fischer, Ion-selective electrodes, *J. Chem. Ed. 56*(6):387-390 (1974).

7. A. E. Greenberg, L. S. Clesceri, and A. D. Eaton, Eds., *Standard Methods for Examination of water and and Wastewater,* 18th ed., American Public Health Association, Washington, DC, 1992.

8. S. E. Manahan, *Environmental Chemistry,* 5th ed., Lewis Publishers, Chelsea, MI, 1991.

9. A. L. Schick, Determination of fluoride in household products by ion selective electrode and Gran's plot, *J. Assoc. Off. Anal. Chem., 56(4)*:798-802 (1973).

10. G. H.Schenk, R. B. Hahn, and A. V. Hartkopf, *Introduction to Analytical Chemistry,* 2nd ed., Allyn and Bacon, Boston, 1981.

11. J. Vesely, D. Weiss, and K. Stulik, *Analysis with Ion-Selective Electrodes*, Wiley, New York, 1978.

12. G. L. Waldbott, Fluoride in food, *Am. J. Clin. Nutr.*, *12*:455-462 (1963).

13. *Toxicological Profile for Fluorides, Hydrogen Fluoride, and Fluorine,* U.S. Department of Health and Human Services, Public Health Service, Agency for Toxic Substances and Disease Registry, Atlanta, GA, July 1991.

Experiment 6

Conductivity of Various Waters

Objective—The objective of this experiment is to show how a transducer, such as a conductivity cell, can be used to determine total dissolved ionic solids in solutions at the part per million level. The variations of specific and equivalent conductance with concentration are also examined.

Introduction—Water in nature contains various amounts of many dissolved substances. These dissolved materials may be gases, solids, and, less commonly, liquids. Undesirable gases, such as hydrogen sulfide, are easily removed by aeration, but dissolved solids are not so easily eliminated. Dissolved substances affect the suitability of water for drinking or other purposes. Federal standards limit the dissolved solids in potable (drinking) water to less than 500 ppm. In industrial applications, such as electroplating, the presence of undesirable dissolved substances can be disastrous to the process and the finished products.

Some environmental waters are highly mineralized. In northeast Florida the amount of dissolved solids in streams varies throughout the year, but it is usually greater (0.95–1.20 g/L) than the recommended standard for potable water. However, many dissolved minerals do not have an adverse effect on human health. Dissolved fluoride, which is naturally present, is beneficial in preventing tooth decay. The principal objection to dissolved solids in water is the scale deposited by the water in taps, water heaters, and boilers.

Theory—In general chemistry three types of solute are discussed: strong and weak electrolytes and nonelectrolytes. Electrolytes conduct electricity readily in aqueous solution, whereas weak electrolytes conduct electricity much less; nonelectrolytes conduct only a negligible amount. Environmental waters contain significant, although small, amounts of dissolved substances, many of which are ionic. Commonly encountered are the halide ions, sulfate, phosphate, sodium, calcium, magnesium, iron, and many others, which are usually present in trace amounts. At a given site, most ion concentrations vary only over a limited range. However, the use of fertilizers, the effluent from sewage outfalls, and the wastes from various industries often magnify the concentrations of many ions above their "background" levels and cause wide variability in their concentrations.

The conduction of electricity in a solution of a strong electrolyte is the same as the conduction of electricity in a solid conductor, such as a copper wire. Thus it obeys Ohm's law,

$$I = E / R \qquad (6\text{-}1)$$

where I is the current in amperes, E is the potential in volts, and R is the resistance in ohms (Ω). The conductivity of a solid (or solution) is a measure of the conductor's capacity to carry an electrical current and, in the case of a solution, is directly related to the concentration of the ions present. The conductivity of a sample of environmental water can be a useful indicator of degraded water quality.

As with a solid conductor, a solution can be characterized by a resistance, the reciprocal of which is the conductance, σ,

$$\sigma = 1/R \tag{6–2}$$

The unit of conductance is ohm^{-1} (Ω^{-1}), which is usually called the mho. The SI unit for ohm^{-1} is the siemen, S, but since the literature and standard test methods usually use mho, this convention will also be used in this manual.

As in a solid conductor, the resistance of an electrolytic solution is proportional to the length, L, of the conductor and is inversely proportional to the cross-sectional area, A, of the electrodes. That is,

$$R = \rho(L/A) \tag{6-3}$$

where ρ is called the specific resistivity, the resistance of a solution of 1 cm length and 1 cm^2 area.

When Eq. 6-3 is substituted into Eq. 6-2, we obtain,

$$\sigma = (1/\rho)(A/L) = \kappa(A/L) \tag{6-4}$$

where κ is termed the specific conductance. Since the specific resistivity has units of ohm-cm, κ has units of mho/cm. It can be defined as the conductance of a conductor 1 cm in length and 1 cm^2 in cross-sectional area. The specific conductance depends on the nature of the conductor (or the nature of the ions in solution), the ion concentration, and temperature.

The measurement of the current-carrying capability of a solution is accomplished with a "conductivity cell." When the cell is filled with an aqueous solution of an electrolyte and an electrical potential is applied across the electrodes, the current developed is found to be proportional to the potential, thus obeying Ohm's law. Alternating current is used to make the measurements to prevent the buildup of ionic charges on the surface of the electrodes (through electrolysis reactions).

Many types of conductivity cell are available, and their sizes as well as their shapes depend on the volume of solution studied and their projected use. A dip-type, shown in Figure 6-1, is convenient for small amounts of solution and also expedites thorough rinsing. The conductance of a solution in such a cell is related to the cell parameters, A and L, through Eq. 6-4. Although theoretically one could calculate the specific conductance from measurements of σ (or R) and values of A and L, it is easier, and more accurate to use a solution of known specific conductance, measure σ, and calculate the ratio A/L. Then the specific conductance of a solution is calculated from

$$\kappa = (\sigma)(\chi) \tag{6-5}$$

where χ is equal to L/A and is known as the cell constant. Typical values of the cell constant are 0.10 and 1.0 cm^{-1}.

Figure 6-1 Dip Type Conductivity Cell

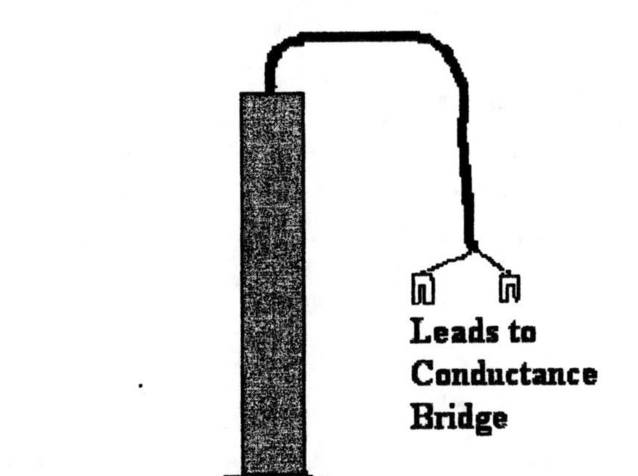

A determination of the cell constant often uses a potassium chloride solution of known specific conductance. The conductance of the solution is measured and the cell constant is calculated using Eq. 6-5. The most frequently used solutions and their specific conductances are given in Table 6-1.

From Eq.6-5 we saw that the units of specific conductance are mho/cm. However, this is rather a large unit, and more often the specific conductance is expressed in terms of millimho/cm or micromho/cm (mmho/cm or μmho/cm). Highly purified water (distilled or deionized) has a specific conductance of 0.5–3 μmho/cm. The specific conductance of potable waters in the United States is 0.05–1.5 mmho/cm. Some industrial wastes may have values exceeding 10,000 μmho/cm (10 mmho/cm).

Although the specific conductance is an important fundamental characteristic of electrolytic solutions, it is not convenient for comparing the relative conductances of solutions. A property based on a molar amount of electrolyte is the equivalent conductance, Λ, the conductance per unit of concentration. Since concentrations are usually expressed in terms of a liter of solution and the specific conductance is the conductance of 1 cm^3 of solution, the equivalent conductance is defined by

$$\Lambda = 1000\, \kappa / C \qquad \textbf{(6-6)}$$

where C is the concentration, in [(moles of ions)÷Z] / L and Λ has units of mho-cm^2 /mol. Some values of the equivalent conductance for potassium chloride are given in Table 6-1. Z is the charge on the ion and accounts for multiply charged ions having the capacity to conduct electricity better than singly charged ions.

Table 6-1 Specific Conductance and Equivalent Conductance of Potassium Chloride Solutions at 25.0°C

Molarity of KCl	κ(μmho/cm)	Λ(mho-cm^2/mol)
0	------	149.9
0.0001	14.9	148.8
0.0010	146.9	146.9
0.01000	1412	141.1
0.1000	12890	128.9

Source: H. S. Harned and B. B. Owen, The Physical Chemistry of Electrolytes, 3rd ed., Reinhold, New York, 1958, pp. 199, 231, and 697.

Unlike the specific conductance, the equivalent conductance increases with dilution, especially for weak electrolytes. As the concentration decreases toward zero, Λ approaches a constant value, Λ^o, the value at infinite dilution. It has been empirically observed that for some electrolytes a plot of Λ versus $C^{1/2}$ gives linear plots in the limit of infinite dilution. Electrolytes that produce such linear plots are defined as strong electrolytes. Plots for weak electrolytes yield curves. This behavior is depicted in Figure 6-2.

Figure 6-2 Comparison of Equivalent Conductances of Strong and Weak Electrolytes

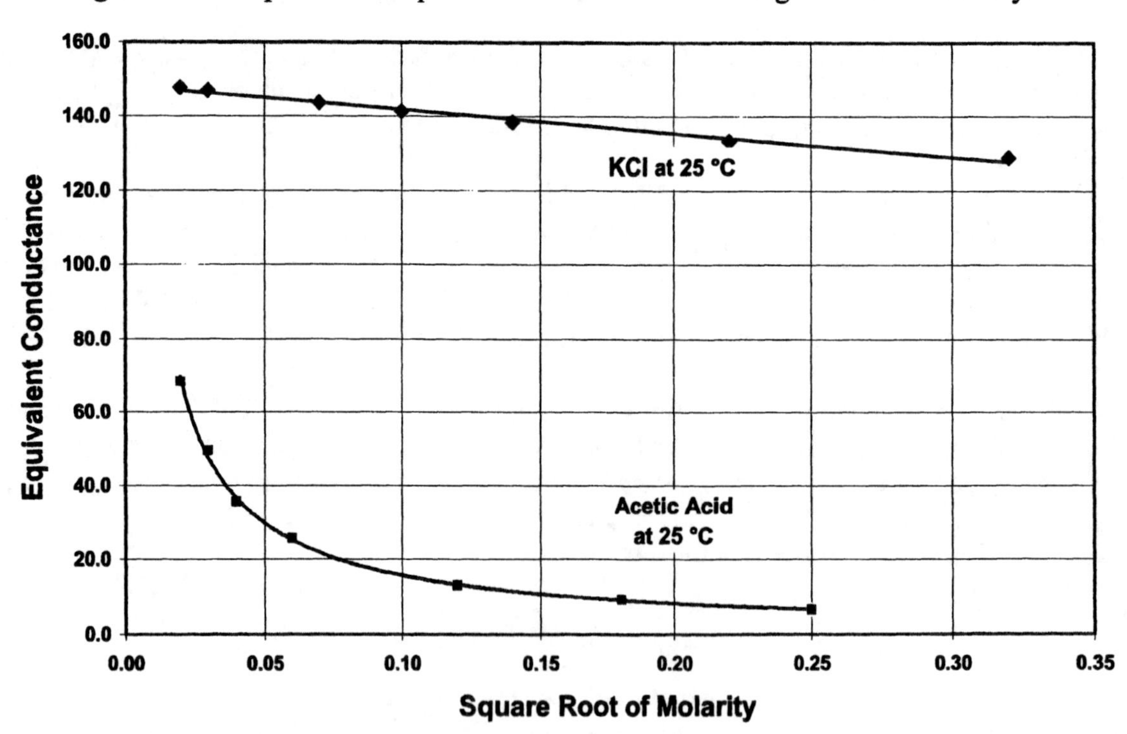

The equivalent conductance can be considered to be a result of the contributions of both the cations and the anions making up the electrolyte. It is possible to calculate the contributions of each type of ion to the total equivalent conductance, at least in the limit of infinite dilution. Thus, for NaCl, $\Lambda^o_{NaCl} = \Lambda^o_{Na^+} + \Lambda^o_{Cl^-}$. Values of separate ionic conductances depend on the nature of the ion and the temperature. Some important values are given in Table 6-2.

Table 6-2 Equivalent Conductances for Cations and Anions, in mho-cm^2 / mol, in Water at 25° C

Ion	Λ^o	Anion	Λ^o
H^+	350	OH^-	198.6
Ca^{2+}	59.5	HCO_3^-	44.5
Mg^{2+}	53.1	CO_3^{2-}	72.0
Na^+	50.1	SO_4^{2-}	80.0
K^+	73.5	Cl^-	76.4
NH_4^+	73.5	F^-	54.4
Ag^+	61.9	NO_3^-	71.4

Source: H. S. Harned and B. B. Owen, The Physical Chemistry of Electrolytes, 3rd ed., Reinhold, New York, 1958, pp. 199, 231, and 697.

Conductivity and Dissolved Solids — Salinity

Table 6-2 illustrates the capacity for different ions to conduct electricity in solution. With the exceptions of hydrogen and hydroxide ions, most ions have equivalent conductances that do not differ to a great extent. If we also consider the fact that for dilute solutions the specific conductance and concentration are linearly related, it is easy to see that estimating dissolved solids can be accomplished by using the specific conductance. Figure 6-3 shows the behavior of the specific conductance of sodium chloride solutions over a limited range of concentration.

The total dissolved solids (TDS) of ionic substances is estimated by multiplying the specific conductance (in μmho/cm) by an empirical factor to obtain ppm TDS. The factor varies from 0.55 to 0.90 (experimentally obtained), where the larger factors are used for saline or boiler water and the smaller factors are used when significant hydroxide or free acid is present. The factor commonly used for environmental samples is 0.564 at 25°C. As an example, consider the data for potassium chloride solutions in Table 6-1. A 0.1000 M KCl solution has a concentration of about (0.1 mol) x (74.5 g/mol) (1000 mg/g) / 1 L = 7450 ppm. Using the specific conductance from Table 6-1 and the 0.564 factor, the TDS is calculated to be 0.564 (12890) = 7270 ppm, a difference of only 2.4%.

When making conductance measurements, it is possible to account for a temperature other than 25°C. A temperature coefficient of 0.019 per degree is applied to all ions except hydrogen and hydroxide, for which the coefficients are 0.0139 and 0.018 per degree, respectively. Equation 6-7 is used to convert specific conductance readings at temperatures other than 25°C to the value expected at 25° C:

$$\kappa(\mu\text{mho/cm}) = \kappa_m / [\,1 + 0.019\,(t - 25)\,] \qquad (6\text{-}7)$$

where κ_m is the measured specific conductance at temperature t (in °C). The effect of temperature on specific conductance must also be taken into account when calculating the cell constant. The equation used can readily be shown to be

$$\chi = (0.001412 / \sigma_{KCl})\,[\,1 + 0.019\,(t - 25)\,]^{-1} \qquad (6\text{-}8)$$

when a 0.0100 M KCl solution is used to calibrate the cell.

Figure 6-3 Specific Conductance of Aqueous NaCl at 25°C

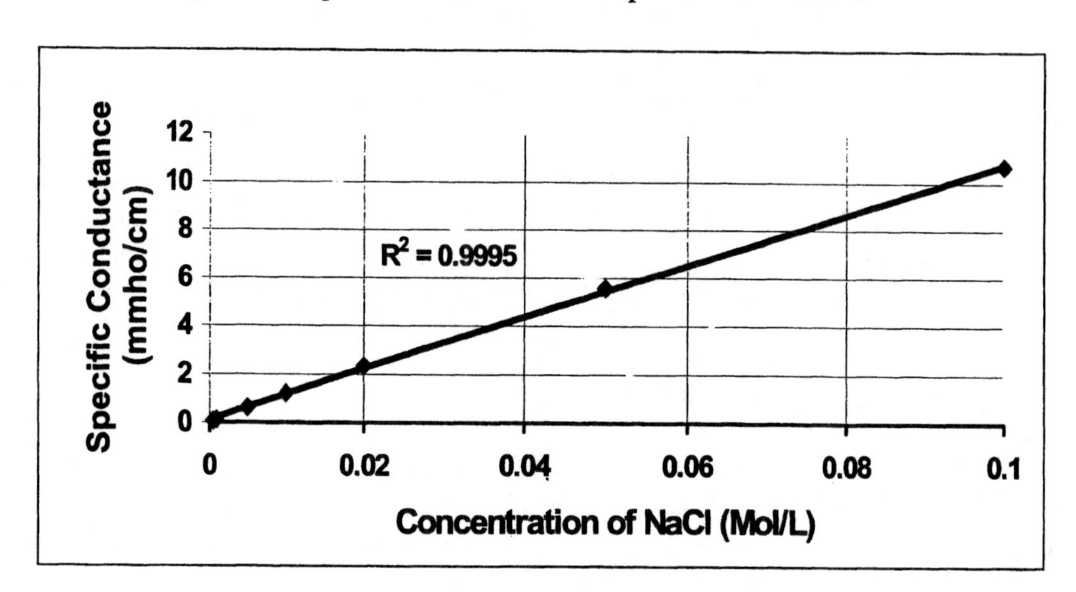

Salinity

A special example of the use and significance of TDS is the **salinity**, an important property of industrial and environmental waters. Salinity is defined as the mass (in grams) of dry salts present in 1 kg of water, seawater, or otherwise. For oceans the salinity varies slightly and has an average value of about 35 g/kg. In percentage this becomes (35/1000) x 100 = 3.5%, or 35 parts per thousand, 35‰.

The salinity is especially significant for estuaries, streams that mix with ocean water. Due to tidal effects and rainfall, estuaries have highly variable salinities. Because many fish use estuaries for spawning and nursery areas, salinities must be appropriate for their survival.

In Experiment 1 density was used to estimate the TDS of river water. The difference in the density between river water and pure water is a measure of the dissolved solids. Refractometry can also be used to determine salinity since it is directly related to the refractive index. However, conductivity measurements give the most precise value of the salinity of all the simple methods.

Safety Precautions

1. Safety glasses must be worn at all times in the chemistry laboratory.

2. If the electrodes of the conductivity cell are replatinized, it is necessary to insulate wire leads with electrical tape. Even a fraction of an ampere can be fatal if you are grounded. Commercial conductivity bridges present no electrical hazards as long as the instrument is intact and not submerged in water—which is very unlikely.

Procedure—There is a variety of conductivity bridges (resistance bridges or Wheatstone bridges) available for making conductivity measurements. Portable bridges are preferred since they can be used in the field for on-site measurements. Various instruments have different capabilities and a number of different readout options. Some instruments have a temperature sensor that greatly facilitates making temperature measurements. Some instruments can give a direct readout of the TDS and others can read salinity directly. The instructor will give specific directions for your particular instrument.

1. Use a 0.0100 M KCl solution (which is provided) to calibrate your conductance cell. Even if the cell constant is known, it should be verified. The black part of the electrode (platinized platinum) must be completely submerged in the solution. Record the temperature of the solution and measure its conductance.

2. Measure the conductance of the following solutions, starting with the most dilute: 10^{-4}, 10^{-3}, 10^{-2}, and 10^{-1} M NaCl (these will be provided). Record the temperature of each solution.

Between samples, shake the cell gently to rid it of most of the previous sample. Rinse the cell with the next test solution, and then briefly swirl or use a magnetic stir bar and plate to mix the next test solution before measuring its conductance.

3. Measure the conductance of five environmental samples, including river water, pond water, well water, tap water, and seawater (or other samples of interest, such as bottled water). Distilled and deionized water are also possible samples for study.

Waste Minimization and Disposal

1. Any waste sodium chloride and potassium chloride solutions can be disposed of in the laboratory drain.

2. The environmental water samples can be used for several subsequent experiments. Dispose of any waste or extra water using the procedure given by your instructor. Unless the water is considered to be hazardous waste, it can be rinsed down the laboratory drain.

Data Analysis

1. Calculate the cell constant for the cell using Eqs. 6-5 and 6-8. How much difference does the temperature make in this calculation?

2. Use the cell constant to calculate the specific conductance (in μmho/cm) for each sample at 25°C, taking temperature into account by using Eq. 6-7.

3. Calculate the TDS (in ppm) for each sample. For the sodium chloride solutions, compare the estimated values with the actual values and calculate the percentage error.

4. Make a table showing sample type, conductance, specific conductance, and TDS.

5. For the sodium chloride solutions, calculate the equivalent conductances using Eq. 6-6. Make a plot (computer generated if possible) of the equivalent conductances (as ordinate) versus the square root of the concentration to test the linearity of the relationship. Report the correlation coefficient and submit the plot.

6. Discuss the results for the total dissolved solids in relation to the location or type of sample measured. If seawater or estuarine samples were studied, calculate and report the salinities and compare with the known value for seawater.

Supplemental Activity—Determine the specific conductances and the equivalent conductances of a weak electrolyte, such as aqueous ammonia. Use a series of five solutions of different concentrations, and plot the equivalent conductance versus the square root of the molarity. Compare with the result for a strong electrolyte (Part 5 in the Data Analysis section).

Questions and Further Thoughts

1. How could conductivity measurements be used to determine the end point in an acid-base titration? Why might this be useful for an environmental sample?

2. Could conductivity measurements be used to determine the end point in a titration in which a complex ion is formed? Explain. How might this be useful in environmental chemistry?

Notes

1. On completing the experiment, the cells should be thoroughly rinsed with DI water. The platinum electrodes have a longer life if they are stored in water.

2. If it is difficult to balance the conductance bridge, it may be necessary to replatinize the electrodes. In this case, see the instructor. Another possibility may be poor contacts between the electrode leads and the terminals of the bridge.

3. If you use one of the older types of bridge that must be balanced using a "Magic Eye," the instructor will provide instructions.

Literature Cited

1. A. E. Greenberg, L. S. Clesceri, and A. D. Eaton, *Standard Methods for the Examination of Water and Wastewater*, 18th ed., American Public Health Association, Washington, DC, 1992.

2. H. S. Harned and B. B Owen, *The Physical Chemistry of Electrolytes,* 3rd ed., Reinhold, New York, 1958, pp. 199, 231, and 697.

3. David R. Lide, Editor-in-Chief, *CRC Handbook of Chemistry and Physics*, 77th ed., CRC Press, Boca Raton, FL, 1996.

4. S. H. Maron and J. B. Lando, *Fundamentals of Physical Chemistry,* Macmillan, New York, 1974.

5. C. N. Reilley and D. T. Sawyer, *Experiments for Instrumental Methods, A Laboratory Manual,* McGraw-Hill, New York, 1961.

Experiment 7

Determination of Chloride Ion in Natural Waters: A Comparison of Methods

Objectives—The objectives of this experiment are:

1. To apply methods introduced in previous experiments to the measurement of an important environmental species, the chloride ion and,

2. To compare the methods in their precision, accuracy, speed, and ease of analysis.

Introduction—In earlier experiments we examined two common but very useful and sensitive transducers: the conductivity cell and the ion-selective electrode. In this experiment results using these instrumental tools are compared with the results from a classical "wet" analytical method, an argentometric (silver nitrate) precipitation titration.

The total concentration of dissolved salts in river waters is much lower than in seawater. It is interesting to note that the principal ions found in river waters are not the same as those found in seawater. The most common anion in river waters is bicarbonate, whereas in seawater it is chloride. In the oceans the most common cation is sodium, Na^+, whereas in rivers it is calcium, Ca^{2+}. Table 7-1 shows the concentrations of various ions in river waters. In Experiment 5, Table 5-1 gives the concentrations of various ions found in seawater.

Table 7-1 Concentrations (in mmol/L) of Common Ions in Various Rivers

Ion	Hudson at Green Island	Mississippi at Baton Rouge	Colorado at Yuma	Amazon at Obides
Na^+	0.21	0.48	5.41	0.07
Mg^{2+}	0.20	0.31	1.24	0.02
Ca^{2+}	0.80	0.85	2.35	0.14
K^+	0.05	0.08	0.11	0.05
Cl^-	0.14	0.43	3.18	0.07
SO_4^{2-}	0.25	0.43	3.01	0.01
HCO_3^-	1.53	1.65	3.00	0.29

The chloride ion is essential in the diet but passes through the digestive tract unchanged to become one of the major constituents in raw sewage. The extensive use of water softeners also adds to chloride levels in sewage and wastewaters. Chloride ion is present in all potable water supplies and, when sodium is also present, results in a salty taste if chloride ion concentrations are greater than 250 ppm.

High chloride levels are not known to have a toxic effect on humans. Chloride ion is essential for human health. Not only is it important because of its role in the electrolyte balance of blood, but it is also essential in the production of stomach acid, HCl.

Large amounts of chloride may have a corrosive effect on metal pipes and may harm growing plants.

The measurement of chloride is important for many reasons. An increase in chloride concentration in milk indicates mastitis in cows. In the processing of foods it is important to know chloride levels.

The measurement of chloride in body perspiration is a useful medical diagnostic procedure. The chloride level in sweat is directly measured on skin by using a flat-tipped chloride ISE. Most chloride abnormalities result in a decreased concentration of chloride. However, increased chloride levels may indicate cystic fibrosis, a childhood lung disease.

There are many methods for determining chloride. The gravimetric method is the most accurate, but it is time-consuming and not well suited to determining low levels of chloride. Two straightforward titrimetric methods based on precipitation with silver nitrate exist, and they differ only in the indicators used for detecting the end point. Another titration method uses conductance measurements to detect the end point. Finally, a potentiometric method, using an ion-selective chloride (solid-state) electrode, is convenient, portable, fast, and accurate to very low concentrations. In this experiment we compare a titrimetric method (using a visual indicator), a titration method using conductance to detect the end point, and finally a method using an ion-selective electrode.

Theory—From basic solubility rules discussed in general chemistry, the only ions that form a precipitate with chloride ion are silver, lead, and mercurous, Ag^+, Pb^{2+}, and Hg_2^{2+}. If a precipitation reaction is to be used as the basis for analysis, the silver ion is preferable since it forms a precipitate with a very small K_{sp} (1.82×10^{-10} at 25°C). Lead chloride, on the other hand, is much more soluble and, like mercury, is very detrimental to the environment. Although silver has its environmental drawbacks as well, silver chloride can be recycled into silver nitrate. Thus, silver nitrate is used to precipitate chloride from aqueous solution according to the equation,

$$Ag^+ (aq) + Cl^- (aq) \longrightarrow AgCl (s)$$

From the titration reaction it is seen that silver and chloride react in a one-to-one ratio, and thus at the equivalence point,

$$\text{mol } Ag^+ = \text{mol } Cl^- \qquad (7\text{-}1)$$

If the chloride is in solution, its molarity can be calculated by changing Eq. 7-1 to the form,

$$\underset{(AgNO_3)}{(\text{mL}/1000)\text{M}} = \underset{(Cl^-)}{(\text{mL}/1000)\text{M}} \qquad (7\text{-}2)$$

using the fact that M = mol/L. Thus, from the titration volume and the molarity of the silver nitrate, the molarity of the chloride can be determined.

Two titration methods for chloride in aquatic samples will be carried out in this experiment. The amount of chloride, in both cases, is determined by measuring the volume of a standard solution of silver nitrate needed to precipitate all the chloride.

Mohr Method

The precise volumetric determination of chloride requires knowing the exact point at which silver nitrate has completely precipitated all of the chloride—that is, the equivalence point. Two methods will be used to detect this point. One of the methods was developed by C. F. Mohr over a century ago. In the Mohr method a solution of potassium chromate is added to a solution of the sample. When this mixture is titrated with silver nitrate, Ag^+ first precipitates Cl^- giving AgCl(s), which is white. This reaction continues until all of the Cl^- has been precipitated. After this point, any more Ag^+ added reacts with chromate to form a pink-brown precipitate of silver chromate. This indicates the complete precipitation of all the chloride. The behavior of the indicator is shown by what is called the "indicator reaction," which in this case is

$$2Ag^+ (aq) + CrO_4^{2-} (aq) \longrightarrow Ag_2CrO_4 \text{ (s, red-brown)}$$

The Mohr method works best in neutral to slightly acidic media. If the pH of the sample is not known, it is determined and the pH brought into the range 6–8 by adding dilute acid or dilute base as required.

Conductometric Titration

Several methods are available for determining the end point without using a visual indicator. These methods usually have the advantage of being faster and more accurate. Some methods are optical, still others are potentiometric, and others are conductometric. In Experiment 6 it was shown that the conductance of a solution varies with the nature of the dissolved ions and their concentrations, and this is the basis for this experiment.

A conductometric titration involves measuring the conductance of the sample solution after successive additions of titrant. The end point is obtained from a plot of the conductance versus titrant volume. Conductometric titration curves have a variety of shapes, depending on the nature of the titrant and the ion under study. Conductometric titration curves are commonly characterized by straight-line segments, with dissimilar slopes on either side of the equivalence point. Figure 7-1 shows the conductometric titration curve for the titration of a strong acid with a strong base.

As the titration proceeds, the highly conducting hydronium ion decreases in concentration up to the end point, after which the highly conducting OH^- causes the conductance to increase rapidly. The intersection of the two lines is the end point.

The relative conductances are determined from equivalent conductances, the values of which are given in Table 6-2. The ionic equation for the titration of chloride with silver nitrate is

$$Ag^+(aq) + NO_3^- (aq) + Na^+ (aq) + Cl^- (aq) \longrightarrow AgCl (s) + Na^+ (aq) + NO_3^- (aq)$$

Before the end point Cl^- is being replaced by NO_3^-. From Table 6-2 it is seen that the nitrate ion has a smaller equivalent conductance than the chloride ion, and the overall conductance of the solution decreases.

After the end point, the addition of excess silver nitrate causes an increase in the conductance.

Figure 7-1 Conductometric Titration Curve for a Strong Acid Titrated with a Strong Base

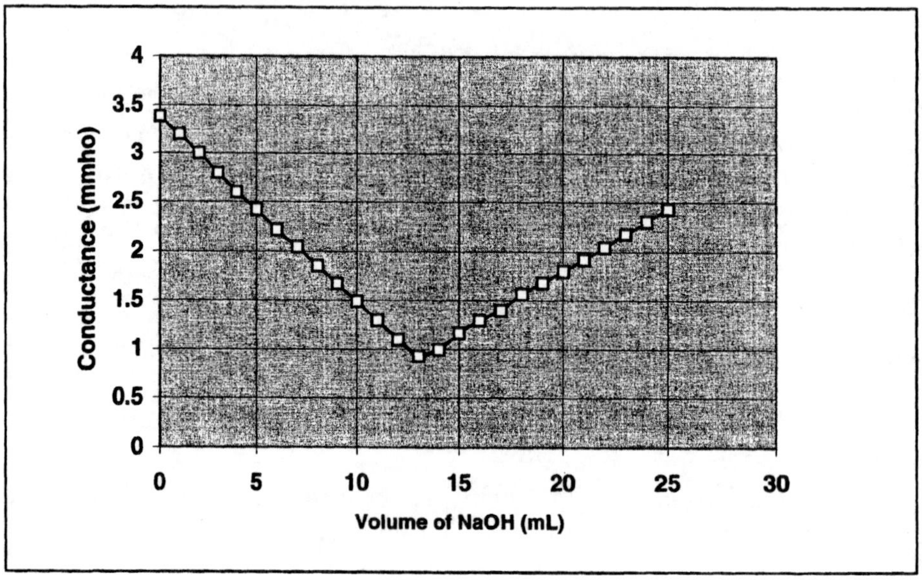

Conductometric titrations are useful for many of the titrations done in chemistry, especially acid-base, precipitation, and complexometric titrations. Neutralization titrations are particularly well suited to this method because of the very high conductances of the hydronium and hydroxide ions.

Ion-Selective Electrodes

Potentiometric measurements are made with an electrochemical cell consisting of electrodes called indicating and reference electrodes. The indicator electrode responds to changes in the concentration of the analyte. The reference electrode is a self-contained system with a constant potential that is in electrical contact with the test solution.

In Experiments 3 and 5 hydronium and potassium ion concentrations were determined using ion-selective glass electrodes, and a solid-state electrode was used to measure the concentration of fluoride ion in Experiment 5. In this experiment, a solid-state electrode will be used to measure chloride ion concentrations.

A solid-state electrode uses a single crystal, or a pellet membrane tip, and may have either an internal Ag-AgCl reference electrode or merely a simple electrical contact to the inner surface of the membrane. The membrane is a conducting solid and contains a lattice ion identical to the anion (or cation) to be measured. Electrical conduction occurs by a lattice defect mechanism in which only the lattice ion (the ion to be measured) moves into vacant positions in the lattice.

ISEs are not completely selective, and ions with similar properties will interfere with an analysis. Thus it is not surprising to find that other halide ions interfere with a chloride analysis using an ISE. The bromide, fluoride, iodide, cyanide, and silver ions seriously interfere with the chloride electrode. Fortunately, these ions are not common in environmental waters. Also, the chloride ISE responds only to free Cl^-. Thus, the presence of complexing metals decreases the measured concentration.

The chloride electrode consists of a silver chloride membrane bonded to an epoxy body. When in contact with a solution containing chloride, a potential that depends on the activity of the chloride ion, increases. The potential developed is measured against an external reference electrode, or an internal reference electrode.

The potential at 25°C for a chloride electrode depends on chloride concentration through the Nernst equation,

$$E = 1.359 - 0.0591 \log[Cl^-] \tag{7-3}$$

based on the half-reaction

$$\tfrac{1}{2}Cl_2(g) + e^- \longrightarrow Cl^-(aq)$$

In Eq. 7-3 it is assumed that the chloride ion concentration is small enough so that the molar concentrations can be used instead of its activity.

Safety Precautions

1. Safety glasses must be worn at all times in the laboratory.

2. Silver nitrate is very toxic if ingested. In contact with the skin, it forms a dark stain as it reacts with the proteins in the skin. The stain is difficult to wash away, but does not cause permanent skin damage. Wear gloves when handling silver nitrate.

3. Avoid all skin contact with the potassium chromate indicator! Chromium(VI) is a carcinogen as well as an acute toxin. Also, chromium(VI) compounds can be explosive in contact with organic matter.

4. If a noncommercial conductivity bridge is used in this experiment, there may be an electrical hazard if there are exposed wires. Exercise great caution in using any electrical apparatus in the laboratory since even a small current can be lethal.

Procedure

A: Standardization of Silver Nitrate

Accurately weigh out about 3.0 g (to 0.1 mg) of silver nitrate in a plastic weighing boat (or a small beaker) and transfer quantitatively to a 250 mL volumetric flask. Dilute to the mark with DI water and stir thoroughly. This yields a solution that is approximately 0.075 M $AgNO_3$. Obtain and record the purity factor (or the assay) from the bottle of silver nitrate. The solution should be stored in a brown bottle in a cabinet to prevent photodecomposition.

B: Analysis by the Mohr Method

1. Pipet 10.00 mL of one of the environmental samples into a 250 mL Erlenmeyer flask, dilute to about 75 mL with DI water, and test with pH test paper. If the pH is not between 6 and 8, use dilute nitric acid or sodium hydroxide, as needed, to bring the pH into the appropriate range.

2. Use a 10 mL graduated cylinder to add 1 mL of 0.25 M potassium chromate to the flask.

3. Rinse and fill your buret with the standard silver nitrate solution. It is advisable to wear gloves to avoid the stain that silver nitrate causes, although these are temporary and not harmful. Titrate the environmental sample with your standard silver nitrate, adding the titrant slowly at first, stirring with a magnetic stir bar and plate. Near the end point the red-brown color of silver chromate persists throughout the yellow suspension. The color change should be faint and should persist for 30 seconds. If the titration volume is less than 10 mL, increase the sample size to make the titration volume 20–30 mL for the second aliquot of the same sample. Do a third titration using the appropriate volume of the same sample.

4. Titrate a blank using the same procedure as was used for a sample, substituting DI water for the sample. (Five drops of a suspension of 0.1 g of calcium carbonate in 2–3 mL water can be used in the blank to emulate the sample matrix in the silver nitrate titration.) The volume of titrant used for the blank is subtracted from the amount of silver nitrate used in the analyses of the samples.

C: Analysis by the Conductometric Method

1. Use the results from the Mohr titration to estimate the volume of the same environmental sample that will require 5–10 mL of titrant to reach the end point. Pipet the sample into a tall form beaker (200 mL) and add enough DI water to completely submerge the electrodes. (The vent hole, if there is one, should also be submerged.) There must be enough space left in the beaker to add two times the volume of titrant needed to reach the end point.

2. Use a magnetic stir bar and plate to continually stir the solution being titrated. Do not stir so rapidly that air bubbles are formed within the solution since they will affect the conductance measurements. After stirring moderately, measure the initial conductance of the solution.

3. Use a buret to add 1 mL increments of silver nitrate, measuring the conductance after each addition. Allow enough stirring so that the titrant completely reacts. Prepare a rough plot of conductance (ordinate) versus volume as the titration is carried out to observe when the end point is reached. An equal volume of titrant should be added after the end point, again in 1 mL increments, measuring the conductance after each addition.

4. Repeat this procedure for two additional aliquots of the **same** sample.

D: Ion-Selective Electrode Method

The instructions for the use of the chloride electrode are given for a combination electrode. This method will be used for the same environmental sample used in Parts B and C for a comparison of the methods, but also several different environmental samples should be measured to compare chloride levels from various locales.

1. Prepare a series of at least four standard solutions, bracketing the concentrations of chloride expected from the previous methods. Table 7-2 gives the compositions of solutions often used to standardize a chloride electrode. Either these standards will be provided, or the instructor will indicate how they are to be prepared.

Table 7-2 The Concentrations of Chloride and Sodium Chloride Used as Standards

Mol/Liter	ppm Chloride Ion	Percentage NaCl
0.0001000	3.545	0.0005844
0.001000	35.45	0.005844
0.01000	354.5	0.05844
0.1000	3545	0.5844

2. Measure 50 mL of sample (standard or test solution) into a 150 mL beaker.

3. Add 1 mL ionic strength adjustor (ISA) and stir thoroughly. The ISA is 5 M $NaNO_3$. Measure the potential to 0.1 mV with the meter in the REL MV mode. Do not stir when making a potential measurement with the combination electrode.

4. Rinse the electrode with DI water, blot dry, and lower into the next solution to be tested. Do not rub the membrane! When measuring standards, test the most dilute standard first, followed by increasingly concentrated solutions (Notes 3–5).

5. Continue the previous procedure with all standards and test samples (preferably four or more samples of different types).

6. Repeat the measurement on one standard solution and one test sample to check reproducibility.

Waste Minimization and Waste Disposal

1. Environmental water samples can be used in several subsequent experiments and should not be disposed of until your instructor so directs. Unless the water samples are considered to be hazardous wastes, they can be disposed of in the laboratory drain (after first getting the approval of your lab instructor).

2. Any unused solid silver nitrate should be put into a container designated by your instructor. Never return material to a stock bottle. Any unused silver nitrate **titrant** should be returned to a designated container. The buret should be rinsed three times with DI water, putting the rinsings into a designated waste receptacle.

3. When a titration has been completed, the titration mixture (water and silver chloride) should be transferred to a waste receptacle. Silver is very toxic and should not be released into the environment. Waste silver chloride can be recycled into silver nitrate. Usually a film of AgCl remains on the glass of the titration flasks and should be removed with small amounts of dilute ammonia. The rinsings should be put into the same waste receptacle as the silver chloride.

Data Analysis—Table 7-3 gives chloride results from the authors' area that may give you an idea of the results expected for an estuary.

Table 7-3 Results of Chloride Analyses for 10 Years on Samples Taken from the St. Johns River, Jacksonville, FL*

Chloride molarity, Mohr method	**0.1584 mol / L**
Range for the Mohr method	**0.0014 mol / L**
Range for the conductometric method	**0.0017 mol / L**
Average salinity (all methods)	**10.16‰**

*The mean chloride molarity over this ten-year period varied from 0.0726 to 0.2601 mol/L.

A: Mohr Method

1. Report the exact concentration of your $AgNO_3$ standard solution to the appropriate number of significant figures.

2. Tabulate the titration volumes, molarities, and sample type. Calculate and report the mean and median concentrations. Also report the range of the values and the relative standard deviation. Comment on the precision of your results. Present your results in a clear manner.

B: Conductometric Method

1. Make plots of conductance (as ordinate) versus volume of titrant, using computer software, if possible.

2. Draw the best straight lines through the points on both sides of the end point, and estimate the equivalence point from the intersection of these lines.

3. Calculate the chloride ion concentrations for each of the three samples, the mean value, and the median.

4. Calculate the range of the three results and their relative standard deviation.

C: The Ion-Selective Electrode Method

1. Make a plot of potential (as ordinate) versus the logarithm of the chloride ion concentration.

2. Carry out a least-squares analysis to determine the slope, the *y*-intercept, and the correlation coefficient

for the standards data. If there is curvature, find the slope, the *y*-intercept, and correlation coefficient for the linear portion of the plot. Compare the experimental slope with the theoretical value from the Nernst equation (Experiments 3 and 5). Use the least-squares results to write the equation for the straight line. Use this equation to calculate chloride ion concentrations from the measured potentials. Prepare a table containing all measured potentials, standard and calculated concentrations, and sample information.

3. To compare the three methods, prepare a table showing the results obtained for the same sample analyzed by all three methods. Include mean values, median values, ranges and relative standard deviations. Discuss any differences in the results and select the "best" method for the chloride analysis, including its limitations.

Supplemental Activity

1. The Fajans (adsorption indicator method) can also be included in this experiment. It gives results comparable to the Mohr method. It is rapid and the end point is sharp. It is carried out in neutral to slightly acidic solution to put the indicator into the correct chemical form.

2. Bottled waters provide interesting samples but generally are low in chloride content.

Questions and Further Thoughts

1. How can interferences such as Br^- or I^- be removed?

2. Saltwater intrusion is becoming very common in many areas of the United States, and chloride determinations can be used to study this phenomenon.

Notes

1. An oxidizing agent called CISA can be used to eliminate some interferences and is available from Orion.

2. On some meters the potentiometer scale is precalibrated to read directly in units such as pAg, pCl, pF, etc. In general, pX $= -\log(a_x)$ where a_x is the activity of the ion of interest.

3. When not in use for short periods of time, the chloride electrode should be stored in 0.01 M NaCl. For longer storage times, drain the electrode, flush the inside with DI water and store dry with the protective cap over the membrane.

4. If test samples have high ionic strengths, the standards should be prepared with a composition similar to that of the samples.

5. If the electrode response is sluggish, the membrane may be coated with deposits. Polish the membrane with the polishing strips provided with the electrode, **and nothing else!**

Literature Cited

1. C. Baird, *Environmental Chemistry,* Freeman, New York, 1995.

2. J. S. Fritz and G. H. Schenk, *Quantitative Analytical Chemistry,* 5th ed., Allyn and Bacon, Boston, 1987.

3. A. E. Greenberg, L. S. Clesceri, and A. D. Eaton, *Standard Methods for the Examination of Water and Wastewater,* 18th ed., American Public Health Association, Washington, DC, 1992.

4. W. C. Pierce, E. T. Sawyer, and E. L. Haenisch, *Quantitative Analysis,* 4th ed., Wiley, New York, 1958.

5. C. N. Reilley and D. T. Sawyer, *Experiments for Instrumental Analysis, A Laboratory Manual,* McGraw-Hill, New York, 1961.

6. G. H. Schenk, R .B. Hahn, and A. V. Hartkopf, *Introduction to Analytical Chemistry,* 2nd ed., Allyn and Bacon, Boston, 1981.

Experiment 8

Determination of the Hardness of Natural Waters—
A: Conventional EDTA Complexometric Titration
B: Commercial Test Kit Determination

Objectives—The objectives of this experiment are:

1. To measure the hardness of environmental waters from several sites to determine the effect of location on hardness

2. To show how hardness may be readily measured in the field using a test kit.

Introduction—The hardness of water was originally defined in terms of its ability to precipitate soap. Calcium and magnesium ions are the principal causes of hardness in water, although iron, aluminum, manganese, strontium, zinc, and hydrogen ions are also capable of producing the same effect. The total hardness of water is now defined as the amount of calcium and magnesium present and is expressed as ppm calcium carbonate.

The procedure for determining both calcium and magnesium, when present together, is found in many schemes of applied analysis, including the analysis of minerals, blood serum, and food, and is the standard method for determining water hardness.

The hardness test is one of the most commonly performed analyses in the water industry. High levels of hardness are undesirable and must be removed before the water is used by the beverage, laundry, metal-finishing, dying and textiles, food, and paper pulp industries. Hardness levels greater than 500 ppm calcium carbonate are undesirable for domestic use and most drinking water supplies average about 250 ppm. Table 8-1 lists the various classes of hardness.

Table 8-1 Classes of Hardness Based on Hardness Range

HARDNESS RANGE (ppm $CaCO_3$)	HARDNESS DESCRIPTION
0–50	Soft
51–150	Moderately Hard
151–300	Hard
>300	Very Hard

Source: G. Tchobanoglous and E. D. Schroeder, *Water Quality*, Addison-Wesley, Reading, MA, 1985.

Theory—Metal ions act as Lewis acids. Anions or molecules with unshared pairs of electrons can act as Lewis bases and covalently bind to metal ions. The electron-pair donors are called ligands, and the species formed in the reaction are known as complex ions if ionic or complexes (or coordination compounds) if neutral. Ligands that bind to the metal at more than one coordination site are called polydentate. The ethylenediaminetetraacetate ion, abbreviated EDTA, is an important polydentate ligand. This species has six donor atoms and is thus hexadentate. It reacts with many metal ions in a 1:1 ratio to form very stable complexes, as in the equation:

$$Co^{3+} + [EDTA]^{4-} \longrightarrow [Co(EDTA)]^{-}$$

EDTA is a tetraprotic acid and is frequently represented as H_4Y. The usual form of EDTA is the disodium salt, Na_2H_2Y. When this form is used as the titrant in a complexometric titration, the titration reaction is:

$$Na_2H_2Y\,(aq) + M^{+2}\,(aq) \longrightarrow MY^{2-}(aq) + 2\,H^{+}(aq) + 2\,Na^{+}(aq)$$

Since hydronium ion is produced, a buffer is necessary since calcium and magnesium ions must be titrated at high pH for stable complexes to be formed and for the proper functioning of the indicator.

The indicators used for EDTA titrations are called metallochromic indicators, and for the most part they are weakly acidic organic dyes. They include Eriochrome Black T (Erio T or EBT), the first one discovered, Calcon, and Calmagite. EBT functions by forming a colored metal complex, $MEBT^{-}$, at the start of the titration. As long as some metal remains unchelated by EDTA, the solution being titrated remains the color of the $MEBT^{-}$ complex. At the equivalence point, EDTA removes the metal ion from the indicator-metal complex by chelating it, and the solution changes color:

$$Na_2H_2Y + MEBT^{-} \longrightarrow HEBT^{2-} + MY^{2-} + H^{+} + 2\,Na^{+}$$

Titrant	color I before end point	color II at end point

The hardness due to calcium and magnesium ions separately can be determined by using the fact that at very high pH, magnesium forms the insoluble hydroxide, $Mg(OH)_2$, whereas calcium remains in solution. The calcium can then be titrated with standard EDTA and its concentration determined. If another sample is titrated with EDTA at a lower pH, both calcium and magnesium ions react. The magnesium hardness is found by the difference in titrant volume used for the two samples.

Some ions, notably iron(III), block the indicator by combining irreversibly with it. In this case the interfering ion must either be removed or chemically tied up before titrating with EDTA.

It is frequently necessary to obtain a quick analysis of a particular chemical on-site. Many test kits have been developed for analysis "in the field." These test kits offer simplicity and portability, in exchange for accuracy and precision. The analysis of well water is often performed to determine whether a water softener would improve water quality.

Safety Issues

1. Safety glasses must be worn at all times in the laboratory and when testing samples in the field.

2. The authors have not had any difficulty with iron masking the indicator. However, if iron concentrations are high enough, an iron precipitate will form that will also interfere in a hardness measurement unless it is filtered.

Procedure

A: Conventional Titrimetry

1. Prepare a standard solution of EDTA by dissolving about 0.9 g (weighed to 0.1 mg) of certified ACS grade Na_2EDTA in enough water to give 250 mL of solution, using a volumetric flask. (For more accuracy, the EDTA can be standardized by titrating samples of primary standard calcium carbonate dissolved in acid, to which a very small amount of magnesium ion is added for the indicator to function properly.)

2. Pipet 10.00 mL of filtered river water sample into a 250 mL Erlenmeyer flask, and dilute to about 50 mL with DI water. Add 15 mL of pH 10 buffer (provided) and mix thoroughly. Add 4 drops of EBT indicator (see Note 2) and titrate with standard 0.01 M EDTA until a pure blue color, with no tinge of purple, is obtained. Repeat this procedure for two additional samples, increasing the volume of sample if not more than 10 mL of titrant is used for the analysis (see Note 3).

3. Repeat Step 2 (triplicate analysis) for another sample type. This sample can be tap water, or any other sample of interest (see Note 3).

B: Using a HACH Test Kit

1. Fill the plastic measuring tube to the top (level) with the water to be tested, and transfer it to the square mixing bottle.

2. Add the contents of one UniVer III Hardness Reagent Powder Pillow, using the clipper to open. Swirl to mix.

3. While swirling the sample and counting drops, add the Titrant Reagent (Hardness 3) dropwise until the color changes from red to blue in one drop.

4. The hardness of the water in grains per gallon, expressed as calcium carbonate, is the number of drops of Titrant Reagent (Hardness 3) used (1 grain = 64.8 mg and 1 gal = 3.7854 L).

5. To convert the result from grains $CaCO_3$ per gallon to mg/L (ppm), multiply the number of drops from Step 3 by 17.1.

Waste Reduction and Disposal

1. Water samples from previous experiments can be used for this experiment. Tap water and bottled water samples can be disposed of by flushing down the laboratory drain. Your instructor will advise you on the disposal of any environmental water samples that may be considered to be hazardous wastes.

2. Any extra solid EDTA taken should be placed in a designated bottle and not returned to the original stock bottle. Any extra EDTA titrant and the products of the titrations can be disposed of in the laboratory drain (if your instructor indicates that this is acceptable practice).

3. There should not be any waste pH 10 buffer, but if there is it can be disposed of in the laboratory drain (again, your instructor will give specific instructions on disposal policy).

Data Analysis

1. Report the concentration of the standard EDTA.

2. Report the total hardness, in ppm $CaCO_3$, for each determination. Since mol EDTA = mol metal from the titration reaction, the moles of calcium carbonate are equal to the moles of EDTA used in a titration. This is finally converted into mg $CaCO_3$/L of sample.

3. Report the mean and median values for each type of sample analyzed.

4. Report the range and the relative standard deviation for each type of sample analyzed in triplicate. Discuss the precision and compare with results obtained in other experiments, such as the determination of chloride.

5. **Compare the two methods**. Compare concentrations obtained using the standard EDTA method with the result using a commercial kit. Discuss the difference, if any.

6. Compare your hardness results with those given in Table 8-1 and classify the hardness of your samples accordingly. Table 8-2 shows hardness results in the authors' area that you can use for comparison.

Table 8-2 Results of Hardness Analyses for Previous Years in the Authors' Area

Average hardness of St. Johns River Water	**1882 ppm $CaCO_3$**
Average range of hardness for St. Johns River Water	**18 ppm $CaCO_3$**
Average hardness of lab tap water	**303 ppm $CaCO_3$**

Supplemental Activity

1. Compare your hardness results with the hardness of seawater. Seawater itself is a good sample to study.

2. If time permits, also determine the hardness of bottled spring water.

Questions and Further Thoughts

1. Could a conductometric titration be used to determine the end point in a complexometric titration, like the one studied in this experiment?

2. Show that the conversion factor for grains calcium carbonate/gal to ppm calcium carbonate is 17.1.

Notes

1. Calcon and some other metallochromic indicators can be used in solid form.

2. Do not use an EBT solution that is older than 2 weeks. Indicator labels should have a date.

3. If the concentration of iron is small, it will form hydroxo complexes that will not interfere with the indicator. If the iron concentration is high, it will precipitate out in the pH 10 buffer. In precipitated form it will still mask the EBT and should be filtered.

Literature Cited

1. A. E. Greenberg, L. S. Clesceri, and A. D. Eaton, Eds., *Standard Methods for the Examination of Water and Wastewater,* 18th ed., American Public Health Association, Washington, DC, 1992.

2. D. A. Skoog, D. M. West, and F. J. Holler, *Fundamentals of Analytical Chemistry,* 7th ed., Saunders, New York, 1996.

3. G. Tchobanoglous and E. D. Schroeder, *Water Quality,* Addison-Wesley, Reading, MA, 1985.

Experiment 9

Spectrophotometry, Colorimetry, and Absorption Spectra: Determining Iron and Manganese in Natural Waters and Sediments

Objectives—The objectives of this experiment are:

1. To illustrate the general principles of absorption spectrophotometry by demonstrating Beer's law

2. To measure the concentration of iron in water and manganese in soil or sediment

Introduction—Colorimetric and spectrophotometric methods are perhaps the most frequently used and important methods of quantitative analysis. These methods are based on the absorption of light by a sample. The amount of radiant energy absorbed is proportional to the concentration of the absorbing material, and by measuring the absorption of radiant energy, it is possible to determine quantitatively the amount of substance present.

Colorimetric and spectrophotometric methods of analysis have been worked out for most of the elements and for many types of organic compounds. Methods based on the absorption of light are well suited to the determination of sample constituents from trace levels up to amounts of 1–2% but are not as frequently used for the analysis of larger (macro) quantities of substances.

Theory—The fundamental law on which colorimetric and spectrophotometric methods are based is the Bouguer-Beer or Lambert-Beer law, usually referred to simply as Beer's law. In mathematical form this law is

$$A = abc \qquad \textbf{(9-1)}$$

where A is the absorbance, a is the absorptivity, b is the internal cell length, and c is the concentration of the solution. When the concentration is expressed in mol/L, Beer's law is written

$$A = \varepsilon bC \qquad \textbf{(9-2)}$$

where ε is called the molar absorptivity, or the extinction coefficient, and C is the molarity. Typically, b is measured in cm, and therefore ε has units of $M^{-1}cm^{-1}$.

The colorimeter, or spectrophotometer, is an important analytical instrument that makes possible a quantitative measurement of the light that passes through a solution. The first step in an analysis is the determination of the optimum wavelength to use for the analysis. The analyte must appreciably absorb light at the wavelength chosen. In a colorimeter exact wavelengths are not used, but rather small bands of wavelengths, and the wavelength chosen for analysis must be such that the absorbance does not change rapidly with the wavelength. If all the wavelengths in this narrow band are absorbed to nearly the same extent, the result is the same as if we isolated a single wavelength to use. Therefore, for an analysis we choose a flat portion of the absorption spectrum (a plot of absorbance versus wavelength). An absorption spectrum is shown in Figure 9-1.

Figure 9-1 Absorption Spectrum of Chromium (III) Nitrate

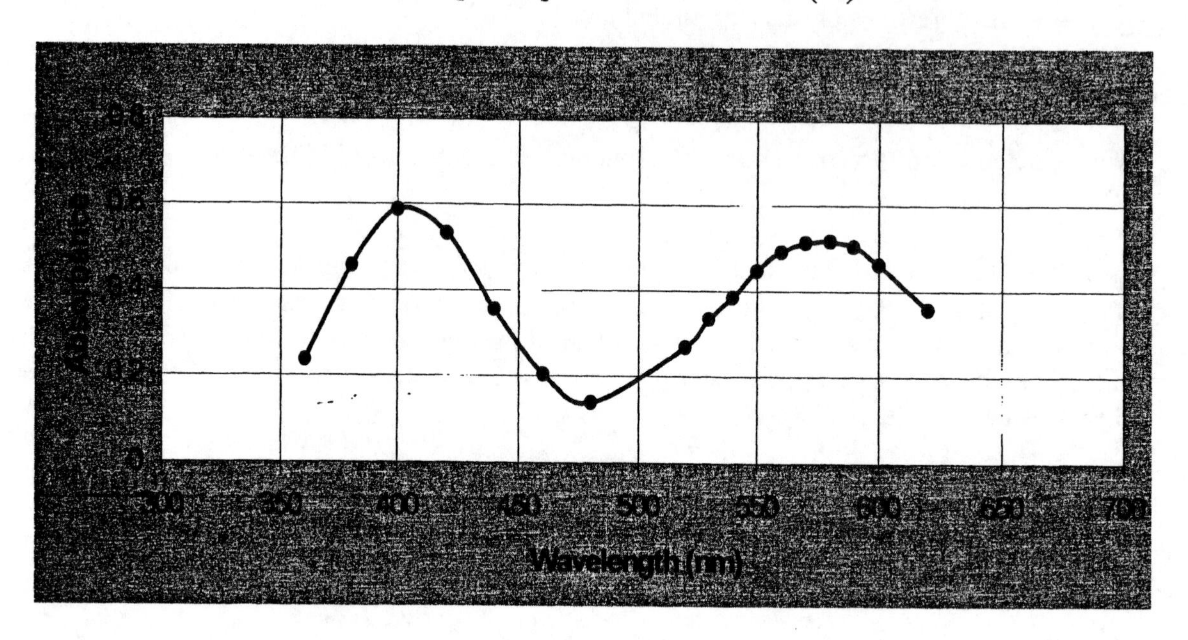

Instruments that measure the absorption of radiant energy, spectrophotometers, have five essential components, as shown in Figure 9-2. For instruments that are used in the visible region of the spectrum, a tungsten filament bulb is used as the source. The **wavelength** of light that enters the system is limited by means of a filter or a monochromator. The **amount** of light that enters the system is controlled with a variable slit or other means. The light then passes through the sample solution held in a glass cell called a cuvette (quartz must be used in the ultraviolet region). Finally, the transmitted light strikes a phototube, or other transducer (such as a photodiode), that converts it into an electric current. The current produced is a function of the radiant power of the light striking the transducer. The current is amplified and is then measured by a meter or a digital readout.

The advantage of a colorimeter is its relatively low cost and simplicity of operation. However, most colorimeters are not able to automatically change wavelength. The output from the source is not constant for all wavelengths, and this necessitates an adjustment in slit width or the sensitivity whenever the wavelength is changed. Also, colorimeters are single-beam instruments and therefore cannot automatically correct for the intensity changes in the light source and variations in detector sensitivity when the wavelength is changed.

The absorbance of a "reagent blank" must be determined at the start of an analysis to correct for any light absorpton by the solvent or reagents.

Figure 9-2 Block Diagram of a Generic Spectrophotometer

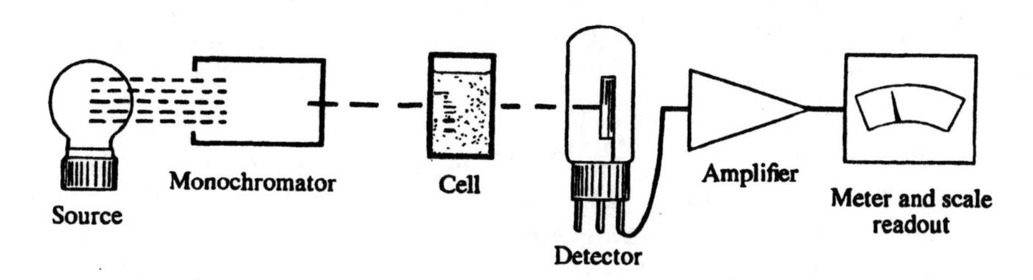

Source: G. H. Schenk, R. B. Hahn, and A. V. Hartkopf, *Introduction to Analytical Chemistry*, 2nd ed., Allyn and Bacon, Boston, 1977. By permission.

Steps in an Analysis

If the analyte is colored, a colorimeter is used for the analysis and the cuvettes can be made of optical glass. If the analyte is not colored, but has an absorption in the ultraviolet, an ultraviolet spectrophotometer is used for the analysis and the cuvettes must be made of quartz or fused silica. In either case, the procedure for an analysis is the same, with the exception of the wavelength region scanned. In the visible region the wavelength range scanned is 760–400 nm, whereas in the ultraviolet region the wavelength range scanned is 400–200 nm.

1. Formation of a Light-Absorbing Species. When a species to be analyzed is not colored and must be analyzed using a colorimeter, it is transformed into a light-absorbing species. (Alternatively, if an ultraviolet spectrophotometer is available and if the species has a functional group that absorbs in the ultraviolet, the use of this instrument may be the easiest way to analyze the sample.) One straightforward way to obtain a colored species is to form a complex. Some metals form highly colored complexes with thiocyanate, SCN^-, for example. A second way to produce a colored species is to transform a metal from a low oxidation state to a higher oxidation state by using an oxidizing agent. Chromium(III), which is only faintly colored, is transformed by oxidizing agents into chromate, CrO_4^{2-} or dichromate, $Cr_2O_7^{2-}$, both of which are intensely colored. Some other types of reactions also produce colored species.

2. Measuring the Absorption Spectrum. The absorbance of the analyte solution is determined as a function of wavelength. The results are plotted (if a recording instrument is not used), preferably using computer software. Ideally, the most intense peak is chosen for the analysis, since it would be the most sensitive to the lowest concentrations. However, if the most intense peak is also sharp, it is better to choose a smaller, broader peak.

The solutions should not contain suspended matter or colloids, which scatter light and distort absorbance measurements.

3. Preparation of a Calibration (Beer's Law) Plot. A series of standard solutions of the analyte is prepared spanning the concentrations expected. The instrument is adjusted to the wavelength chosen for the analysis, and the absorbance of each standard is measured. A plot of absorbance (ordinate) versus concentration is made, preferably using a computer. A least-squares analysis is carried out to obtain the equation of a straight line from which solution concentrations are calculated from measured absorbances. The correlation coefficient from this analysis indicates the precision of the results. Curvature of the plot may indicate a change of equilibrium position of the analyte species with dilution and may have to be taken into account. A Beer's law plot is illustrated in Figure 9-3.

4. Measuring the Sample. The absorbance of the sample is measured at the wavelength used for the calibration. The concentration of the analyte is found from the Beer's law plot (either by estimating directly from the plot or by calculation using the straight-line Beer's law equation). The analyte concentration should be between the extreme limits of the plot; if not, its concentration or the concentrations of the standards should be adjusted accordingly.

Figure 9-3 Beer's Law Plot for Permanganate at 525 nm

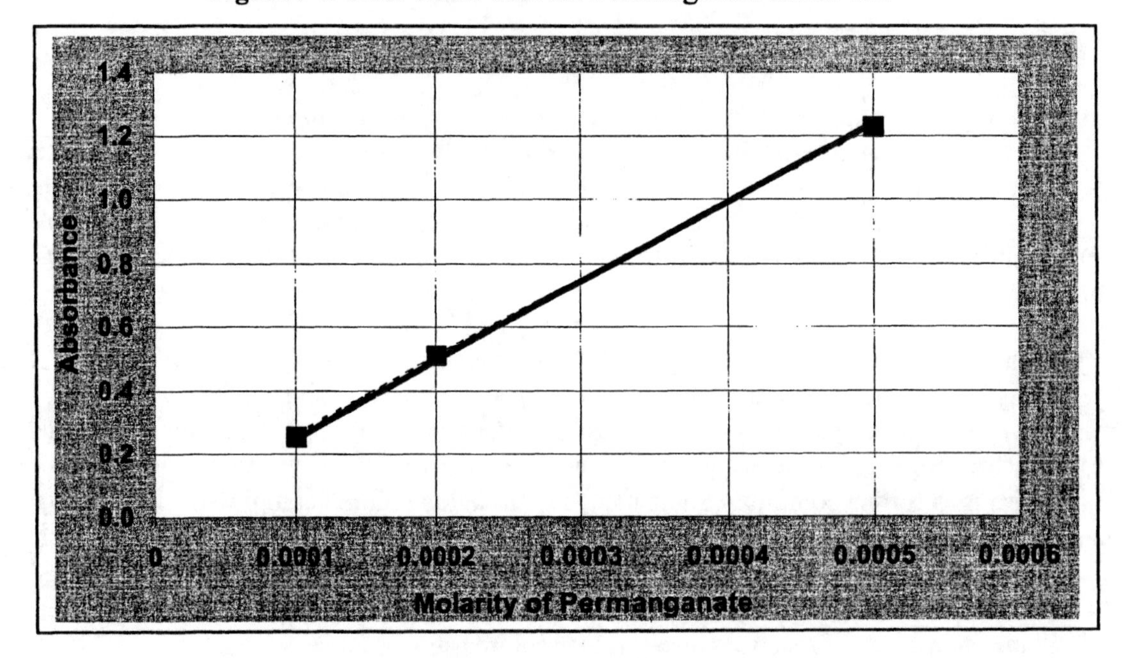

Part A: Determining Iron in Natural Waters

Introduction—Iron is found throughout the environment, often in large amounts. It enters the hydrosphere through the weathering of iron salts and minerals. Both iron(II) and iron(III) are found dissolved in water, often in colloidal form, or as inorganic and organic iron complexes. There are many industrial sources of iron, including canneries, tanneries, textile mills, shipping, and metal-cleaning operations.

Large concentrations of iron discharged into a stream or lake may have deleterious effects on aquatic life. A limit of 0.3 mg/L of iron is recommended for food and dairy product processing, soft drink manufacture, and brewing, mainly because of taste.

Iron is a vital element in the respiratory processes of many animals, including humans. The human body has a great demand for iron, and 4 grams are found in the average human. Iron-containing proteins transport oxygen, catalyze the decomposition of peroxides, and play an essential role in the body's energy-generating processes. It is possible to ingest too much iron, which may cause liver damage. Thus, iron vitamin supplements contain cautionary statements.

Theory—A simple but sensitive procedure for the colorimetric determination of iron entails chelating ferrous iron with three molecules of 1,10-phenanthroline (phen) in a solution buffered at low pH,

$$\mathbf{Fe^{2+} + 3\,phen \longrightarrow [\,Fe(phen)_3]^{2+}}$$

The orange-red complex has an absorption maximum at 510 nm.

A preliminary acid digestion of the sample is carried out in a fume hood to destroy organic matter and also to remove cyanide and nitrate which interfere with the analysis. Hydroxylamine hydrochloride is then added to reduce all iron(III) to iron(II), which is the effective complexing species. Then an excess of phen is added to the sample at a pH between 3.5 and 4.5. The low pH prevents other metals from precipitating and provides rapid reaction and color development. The concentration range of this method is 0.025–3.0 mg/L. Concentrations greater than 3.0 mg/L can be determined after diluting.

An advantage of the 1,10-phenanthroline method is its use of slightly acidic media. This prevents not only the precipitation of hydroxides, but also the phosphates and other anions of many metals. It is necessary that iron be present in a form that reacts completely with 1,10-phenanthroline in a reasonable period of time. This means that iron must not be bound to pyrophosphates or other ligands that form stable complexes; also, phosphate precipitates that contain iron must be prevented from forming. Therefore, the usual procedures in which sodium acetate is used to adjust the pH to 3.5–4.5 are not adequate for biological samples due to the possibility of precipitating ferric and aluminum phosphates. This is avoided by using sodium citrate.

Safety Precautions

1. Safety glasses must be worn at all times in the chemistry laboratory.

2. Use gloves when handling environmental samples.

3. Use extreme care in handling concentrated sulfuric acid. **Wear gloves and goggles.**

Procedure

A: Preliminary Work

1. Attempt to form a complex between thiocyanate ion and iron(II) by mixing together 5 mL each of 0.01 M KSCN and 0.01 M $Fe(NO_3)_2$ (these are provided).

2. Repeat Step 1, this time using 0.01 M $Fe(NO_3)_3$ in acid (also provided) in place of 0.01 M ferrous ion.

3. Use a colorimeter or a recording spectrophotometer to measure the absorption spectrum of the solution prepared in Step 2, scanning from 760 to 400 nm. If necessary dilute the solution to bring its absorbance within the range of the instrument. Use water as the reference.

4. If a Spectronic 20 is used in Step 3, the instrument must first be adjusted to 0%T, at any wavelength. With no sample in the cell compartment, switch the instrument to the Transmittance Mode (if this type of instrument is available; if not, read the %T scale for the following adjustment). Adjust to zero with the front, left-hand knob. Then place a cuvette with water into the cell compartment and use the front, right-hand knob to adjust the instrument to 100%T. This last adjustment must be made each time the wavelength is changed since the intensity of light from the source and the sensitivity of the phototube vary with wavelength.

5. After the adjustments of Step 4, the sample is placed in the cell compartment and its absorbance is read (change the mode to absorbance, if appropriate). If a digital readout instrument is not used, you may wish to make $\%T$ readings and convert them to absorbance via $A = 2 - \log(\%T)$. The rationale for this is that the $\%T$ scale is linear and the absorbance scale is logarithmic, and less error is made in reading the linear scale.

6. Repeat Steps 1, 3, and 4 using Fe(II) nitrate and 5 mL of 0.3% 1,10-phenanthroline solution instead of thiocyanate.

B: Procedure for Iron in Water

1. **Preparation of Standard 100 ppm Iron Solution (this may be provided)** Weigh 351 mg of high quality ferrous ammonium sulfate hexahydrate, $FeSO_4 \cdot (NH_4)_2SO_4 \cdot 6H_2O$, and quantitatively transfer to a 500 mL volumetric flask. Add 50 mL DI water followed by 1 mL of concentrated sulfuric acid. Dilute to the mark with DI water and mix thoroughly.

2. **Preparation of Standard Solutions** Prepare four standard solutions of iron(II) having the following concentrations: 0.5, 1.0, 2.0, and 5.0 ppm. Pipet 0.5, 1.0, 2.0, and 5.0 mL of 100 ppm stock solution into 100 mL volumetric flasks and dilute to the mark with DI water.

3. **Obtaining a Beer's Law Plot** Transfer a 5 mL aliquot of the 0.5 ppm iron standard to a 125 mL Erlenmeyer flask and test the pH with test paper. If greater than 4.5, add enough 0.2 M sulfuric acid dropwise to bring the pH to about 3.5, counting the number of drops. Again counting drops, add sodium citrate (259 g/L) buffer to bring the pH to about 4.5. Pipet 1 mL of 10% hydroxylamine hydrochloride and 3 mL of 1,10-phenanthroline into the sample, mix, and allow 5 minutes for color development.

Use the same number of drops of sulfuric acid and sodium citrate for the remaining three standard solutions, followed by 3 mL of 0.3% 1,10-phenanthroline and 1 mL of 10% hydroxylamine. Mix well.

After adjusting the 0 and 100%T on the colorimeter at 512 nm, use water as the reference and measure the absorbance of each standard. (A reagent blank can also be used if desired. This consists of all substances added to the sample, is treated the same way as the sample, and accounts for any absorbance due to these materials.)

4. **Analysis of Samples** Natural and tap water samples often have less than 0.5 ppm iron. A water faucet that has not been used for some time may furnish a good sample for iron analysis. Some well waters are high in iron content as well. However, even very dilute samples are within the range of this experiment. Determine the iron in several environmental water samples.

Treat samples the same way as the standards, adding sulfuric acid initially, if necessary, followed by citrate buffer, reducing agent and the indicator. Use 5 mL samples and adjust the pH for each sample individually.

Waste Minimization and Disposal

1. Water samples, if they are considered to be hazardous wastes, are disposed of according to the directions of your instructor. Otherwise, they may be flushed down the laboratory drain.

2. Waste ferrous and ferric nitrates can be flushed down the laboratory drain. Follow the instructor's directions for disposing of waste potassium thiocyanate. Do not flush it down the laboratory drain!

Data Analysis

1. Submit the two absorption spectra. From the iron-phenanthroline absorption spectrum decide what wavelength(s) can be used for analysis. Which particular peak, if there is more than one, would be best for an iron analysis?

2. Briefly discuss the effect of the ligand on the wavelength and the maximum absorbance of the peaks.

3. Calculate the extinction coefficient for the largest peak in the absorption spectrum of the iron-phenanthroline complex. (An excess of 1,10-phenanthroline was used.) Assume a cell path length of exactly 1.0 cm. What quantitative information does this provide?

4. Prepare a Beer's law plot for the standard solutions (absorbance as ordinate versus concentration as abscissa). Carry out a least-squares analysis and determine the slope, the *y*-intercept, and the correlation coefficient for the best straight line; comment on the linearity of the plot.

5. Use your Beer's law plot to determine the concentration (in mg/L) of iron in each sample studied. The concentration of a sample can be obtained directly from the Beer's law plot. Alternatively, the equation of a

straight line can be obtained from the slope and *y*-intercept and the concentration calculated from the sample's absorbance.

If duplicate determinations were done, report average values and the individual values of concentration and discuss the reproducibility.

6. Discuss the magnitude of the iron levels with respect to the sampling site.

7. For the instrument you used, what would you estimate to be the lowest concentration of iron that would be reliable?

Questions and Further Thoughts

1. Where are the major deposits of iron ore in (a) the United States, (b) the Western Hemisphere?

2. Why are elevated levels of iron toxic to humans? What is its major physiological effect?

3. Why is SCN^- not as good as 1,10-phenanthroline in a colorimetric analysis of iron?

Notes

1. The instructor should supervise the handling of concentrated sulfuric acid. A more dilute sulfuric acid solution can be used in place of the concentrated acid.

2. The absorption spectra are best measured with a recording instrument, if one is available.

3. If a Spectronic 20 is used in this experiment, allow at least 30 minutes warm-up time. Keep the instrument on for the entire laboratory period.

Literature Cited

1. J. S. Fritz and G. H. Schenk, *Quantitative Analytical Chemistry,* 5th ed., 331-359 Boston, 1987, pp. 331-359.

2. A. E. Greenberg, L. S. Clesceri, and A. D. Eaton, Eds, *Standard Methods for the Examination of Water and Wastewater,* 18th ed., *American Public Health Association,* Washington, DC, 1992.

3. W. L. Masterton and C. N. Hurley, *Chemistry Principles & Reactions,* 3rd ed., Saunders New York, 1997.

4. D. A. Skoog, D. M. West, and F. J. Holler, *Fundamentals of Analytical Chemistry,7th* ed., Saunders, New York, 1996.

Part B: Manganese

Objective—The objectives of this experiment are to illustrate how serious interferences, such as chloride ion, can be eliminated by simple chemical means, and how oxidation can be used as a second method of color development in colorimetry. We also expand the previous part of this experiment to include solid (that is, soil and/or sediment) samples in the colorimetric technique.

Introduction—In Part A of this experiment we saw how colored species could be formed by complexation. However, several elements can be oxidized from lower to higher oxidation states that absorb in the visible region of the electromagnetic spectrum. Lower oxidation states of transition elements (for example, Cr^{3+}) are often only faintly colored, whereas higher oxidation states of the same element (CrO_4^{2-} or $Cr_2O_7^{2-}$) are often intensely colored. In this experiment we oxidize manganese(II), which is only faintly colored, into permanganate, MnO_4^-, which is intensely colored.

Most environmental samples contain a wide variety of substances, and when one substance is to be determined, such as manganese, the presence of interferences must be considered. The determination of manganese depends on oxidizing Mn^{2+} to manganese(VII), and if organic matter and/or chloride ion are present, oxidizing agents will be consumed as organic matter is oxidized to CO_2 and chloride is oxidized to chlorine gas. In the presence of a substantial amount of chloride, after permanganate is formed its color will fade as it oxidizes Cl^- to Cl_2. As we shall see, both chloride and organic matter can be chemically eliminated.

Manganese in the Environment

Although manganese is not an extremely important pollutant, it is ubiquitous in soils, air, and water. The level of manganese can be quite large in some locales as natural "background." In areas where it is used in manufacturing or is present in landfills or sewage, it can reach levels that may be toxic to plant and animal life.

Manganese is present to about 1 ppt in the earth's crust and is a component of many minerals. Manganese occurs naturally in almost all soils, and the average background level varies from 40 to 900 ppm, with a mean of around 330 ppm.

A survey of surface waters in the United States in the 1960s detected dissolved manganese at about 50 ppb, with a range of 0.3–3230 ppb. In general, drinking water contains about 4 ppb manganese. The concentration of manganese in seawater varies from about 0.4 to 10 ppb.

Manganese in water may become highly bioconcentrated, especially at lower trophic levels (at the lower end of the food chain). The bioconcentration factor, BCF (the ratio of the concentration of a substance in an organism relative to that in the organism's surroundings) for phytoplankton is 2500–6300 and 35–930 for coastal fish. For marine and freshwater plants the BCF is 10000–20000, whereas it is 10000–40000 for invertebrates. However, biomagnification of manganese is not significant.

Manganese levels in the atmosphere are about 20 ng/m^3. A 1980s study detected manganese concentrations of 33 ng/m^3 in urban air, but only 5 ng/m^3 for nonurban air. The half-life of airborne particles containing manganese is of the order of days.

Manganese enters the environment when fossil fuels are burned and when leachate from landfills and stormwater runoff from terrestrial soils runs into streams. Manganese compounds are used in the manufacture of batteries, pesticides, fertilizers, dietary supplements, and ceramics. The metal itself is used to produce high-quality steels. (Manganese, in the form of methylcyclopentadienyl manganese tricarbonyl, MMT, has replaced lead as an "anti-knock" additive in gasoline in some countries. MMT has been adopted in Canada and some European countries, but not as yet in the United States.)

In the environment manganese is found primarily as salts or oxides of manganese(II) or manganese(IV). Manganese(II) is the predominant form of manganese at pH 4–7, and carbonate ion is the chief anion associated with manganese (II) in water. The low solubility of $MnCO_3$ in environmental waters (about 12 ppm at 25°C and pH 8) limits the concentration of Mn^{2+}. Manganese is often transported in streams on suspended silt that eventually settles out.

There is evidence that manganese is needed in the human diet. A deficiency of manganese can lead to convulsive disorders. It is also required for good health, reproduction, and the proper growth of other animals. A deficiency of manganese results in skeletal abnormalities. The basic role of manganese appears to be in the functioning of certain enzymes.

The recommended daily intake of manganese for adults is 2.5–5.0 mg. Food is the main source of manganese for humans, and the manganese content of a normal diet is 2–9 mg per day. In foods the highest levels are found in nuts and grains, and lower levels are found in meats, milk, fish, and eggs. A large excess of manganese can be harmful. Table 9-1 lists manganese levels found in various tissues, organs, and foods.

Theory—There are several colorimetric methods for determining manganese. The persulfate procedure is preferred for environmental samples because interferences can be eliminated in this method. Mercuric ion is used to control the interference from chloride, which is present in many environmental samples.

Although persulfate is able to oxidize manganese(II) to permanganate, MnO_4^-, the reaction is slow. Thus, the reaction is carried out at an elevated temperature and in the presence of silver ion, which acts as a catalyst. The balanced equation for the production of the colored species is

$$2\,Mn^{2+}(aq) + 5\,S_2O_8^{2-}(aq) + 8\,H_2O(l) \longrightarrow 2\,MnO_4^-(aq) + 10\,SO_4^{2-}(aq) + 16\,H^+(aq)$$

The color developed is stable for 24 hours or longer when excess persulfate is present and in the absence of organic matter. Organic matter can be destroyed by prior digestion with nitric acid.

Interferences As much as 200 mg of chloride per 100 mL of sample can be kept from interfering by adding 2 g of $HgSO_4$, which reacts to form only slightly dissociated chloride complexes. Bromide and iodide will still interfere, and only trace amounts are allowed.

If only small amounts of organic matter are present the persulfate procedure can be used if the heating time is increased after more persulfate has been added. However, if higher levels of organic matter are present, an initial digestion with nitric acid is necessary. If high levels of chloride are present, boiling with concentrated nitric acid aids in its removal.

If the original sample is in an oxidizing environment, low results may occur due to the precipitation of MnO_2. This is avoided by adding hydrogen peroxide to oxidize manganese (IV) back to permanganate,

$$2\ MnO_2(s) + 3\ H_2O_2(aq) \longrightarrow 2\ MnO_4^-(aq) + 2\ H^+(aq) + 2\ H_2O(l)$$

All the necessary reagents for eliminating interferences in the manganese analysis, after digestion, are provided together in what is termed the "special reagent," the composition of which is given in Note 2.

Table 9-1 Manganese Levels in Tissues, Organs and Foods

Sample Type	Humans (ppm)	Rats and Rabbits (ppm)
Liver	1.2–1.7	2.1–2.9
Kidney	0.6–0.9	0.9–1.2
Muscle	0.09	0.13
Fat	0.07	—
Bone	0.06	—

Food Type	Level(ppm)
Nuts	18–47
Grains	0.4–41
Legumes	2.2–6.7
Fruits	0.2–10.4
Vegetables	0.4–6.8
Meats, Fish, Milk, Eggs	0.02–0.5

Source: H. Fore and R. A. Morton, Manganese in rabbit tissues, *J. Biochem.*, 51:600-603 (1952);
J. A. T. Pennington et al., Mineral content of foods and total diets: the selected minerals in foods survey, 1982 to 1984, *J. Am. Diet Assoc.*, 86:876-891 (1986);
G. L. Rehnberg et al., Chronic ingestion of Mn_3O_4 by rats: tissue accumulation and distribution of manganese in two generations, *J. Toxical. Environ. Health*, 9:175-188 (1982);
K.Sumino et al., Heavy metals in normal Japanese tissues: amounts of 15 heavy metals in 30 subjects, *Arch. Environ. Health*, 30:487-494 (1975);
I. H. Tipton and M. J. Cook, Trace elements in human tissue, Part II, adult subjects from the United States, *Health Phys.*, 9:103-145 (1963).

Safety Precautions

1. Nitrogen dioxide is extremely toxic and you should avoid breathing in any of its vapors. *The Merck Index* (see Ref. 2) describes it as a "deadly poison!" A 100 ppm level is dangerous for even a short exposure, "and 200 ppm may be fatal."

2. Permanganate, hydrogen peroxide, ammonium persulfate, and nitric acid are all strong oxidizing agents, and care must be exercised in handling these substances. Gloves and safety glasses are mandatory. They must also be kept from coming into contact with organic matter.

3. Silver nitrate and mercuric sulfate are extremely toxic. Exercise care in handling the special reagent which contains these two compounds.

4. Do not dispose of samples by pouring down the drain. Follow the instructions of your lab supervisor in disposing of materials used in this experiment. Detoxification of both silver and mercury can be accomplished by precipitating their highly insoluble sulfides. Refer to the Waste Minimization and Disposal section about additional safety factors.

Procedure

1. Accurately weigh (to 0.1 mg) a sample of about 10 g of soil or sediment into a tared 150 mL beaker, after decanting off excess water (if necessary). The sample may be dried prior to this step, if desired, or a dry weight can be determined as described in Supplemental Activity 3. Alternatively, the samples that were sectioned and ashed in Experiment 2 can be used for this experiment.

2. Repeat Step 1 for two additional, different environmental samples and prepare a blank at the same time. The blank contains every reagent used for the samples and is treated exactly the same way as the sample. This is called a reagent blank. (Your instructor may also want you to take two aliquots of the same sample to determine the reproducibility.)

3. Working in a fume hood, digest the samples by adding 10 mL DI water followed by 10 mL of high-purity nitric acid (see Safety Precautions 1 about the danger of NO_2 fumes that form when nitric acid decomposes). Cover the beaker with a watchglass and heat to just below the boiling point for 20–30 min. Additional acid and water are added, if needed. (While waiting for digestion, start on Step 11).

4. Remove the beaker from the hotplate using crucible tongs and add 10 mL DI water. Cool the sample to near room temperature, using tap water or an ice bath around the beaker, to save time.

5. Use DI water to wet a circle of medium porosity filter paper (for example, Whatman Number 1) in a funnel. Filter the sample into a 100 mL volumetric flask. Rinse the residue in the original beaker with several small portions of DI water, adding the rinsings to the funnel. Finally, rinse the funnel a few times with small portions of water.

6. Add DI water to the volumetric flask to bring the volume to 100 mL, stopper, and mix thoroughly. Note the color of the solution.

7. In a hood, pipet a 10 mL sample of the extract into a labeled 125 mL Erlenmeyer flask. Add 5 mL of special reagent (see Note 2) and 2 drops of 30% hydrogen peroxide (see Note 3). Boil for 5 min, add 1 g of ammonium persulfate, and boil for 1 more minute.

8. Remove the flask from the hot plate, let it stand for 1 minute and then cool with tap water to near room temperature. The pink-violet color of permanganate should be evident. Treat each sample and the blank in an identical manner. (It is important to boil all samples at the same rate and for an identical period of time.)

9. Quantitatively transfer the sample to a 25 mL volumetric flask, rinsing the Erlenmeyer flask two or three times with small portions of DI water. Fill to the mark and mix thoroughly.

10. Measure the absorbance of the previous solution at 525 nm, using the blank you prepared as the reference. If the measured absorbance is too great, dilute an appropriate-sized aliquot in a volumetric flask to bring it within range.

11. Prepare four to five manganese standards while waiting for the digestion, cooling and filtering in Steps 3–5. These should contain 1–10 ppm manganese. Use a 100 ppm stock solution of $KMnO_4$ to make the standards (this may be provided). Measure the absorbance of each standard at 525 nm, using DI water as the reference.

Waste Minimization and Disposal

1. Sediment or soil samples may be used in other experiments and should not be discarded except with the approval of your instructor.

2. The dilute standard potassium permanganate solutions should not be disposed of in the laboratory drain. Permanganate is readily detoxified by organic matter to produce manganese(II), which is common and harmless in the environment at low concentrations.

3. The digested samples cannot be disposed of by flushing down the laboratory drain. These samples contain silver ion, mercury (II), persulfate, and nitric acid, and wastes must be put into a designated container for later commercial waste disposal.

Data Analysis

1. Prepare a Beer's law plot for permanganate, preferably using computer software. Do a least-squares analysis on the calibration data to obtain the slope, y-intercept, and correlation coefficient for the straight line. Write the equation for the straight line obtained.

2. Determine the concentration of manganese in the diluted sample solutions either by approximation from the Beer's law plot or by calculation using the measured absorbances and the equation for the straight line. Calculate the amount of manganese in the diluted sample solutions by multiplying the concentration from Beer's law times the volume of the measured solution (25.00×10^{-3} L). To obtain the manganese in the original extract, the previous quantity must be multiplied by 10 (since 10 mL of the original 100 mL was

used for analysis). Finally, convert this result to mg manganese per kg of sample by dividing the mg of manganese by the original mass (in kg) of sample used.

3. Discuss the difference in manganese levels for the different environmental samples on the basis of the location of the sampling sites.

4. What are possible experimental errors in this experiment? What is the lowest level of manganese that could be detected by this method?

5. If a sample was analyzed in duplicate, discuss the reproducibility.

6. Discuss the significance of the value of the *y*-intercept obtained for the Beer's law plot.

Supplemental Activity

1. The manganese levels can be confirmed by analyzing the same sample extracts using atomic absorption. The soil and sediment sample extracts obtained in this experiment can be aspirated directly, and the same standards used for colorimetry can be used for the atomic absorption analysis.

2. Iron in soils and sediments can also be determined colorimetrically (or by atomic absorption), using the procedure given in Part A of this experiment for the extracts. A comparison of the iron and manganese levels for the samples can then be made.

3. Often, the results of an analysis of a solid sample are expressed as mg of analyte per kg of dry weight. This can be readily done by using a 1 g sample of wet sediment and drying it to a constant weight at 110°C. Then the corresponding dry weight of the sample is: (wet weight measured out for analysis)(dry weight of sample/wet weight of sample). The ratio (dry weight of sample/wet weight of sample) is simply the fraction of the sample that is dry weight. Alternatively, the results from drying the samples in Experiment 2 can be used for converting wet weights to dry weights, assuming the same percentage of water in both samples.

4. The core sections from Experiment 2 can be used for the manganese analysis using the same procedure as described for sediments. The results can then be used to give a manganese profile.

Questions and Further Thoughts

1. Manganese levels in the environment vary considerably. Some vegetation has very elevated manganese levels (swamp grass, for example) and decayed plant matter can enrich the soil in manganese.

2. One of the authors has found that some black liquor samples (from paper manufacture) are very high in manganese, and these provide good samples to study.

3. Sandy samples are usually low in manganese (as well as most other elements). Why?

4. Another interesting study is determining manganese in seashells.

Notes

1. Ordinary nitric acid contains a variety of metals at high levels. The result of using impure nitric acid can be readily seen by preparing a reagent blank using ordinary nitric acid. The result of the analysis can be compared with the assay on the bottle.

2. The special reagent contains mercury(II) sulfate, phosphoric acid, silver nitrate, and nitric acid. It is prepared by dissolving 75 g of $HgSO_4$ in 400 mL of concentrated HNO_3 and 200 mL DI water. To this mixture add 200 mL of 85% H_3PO_4 and 35 mg of $AgNO_3$ and dilute to 1 L.

3. Hydrogen peroxide should be refrigerated.

4. Your lab instructor may want the manganese standards to be treated the same as the samples to provide a similar matrix. In this case the standards can be made by dissolving manganese metal in concentrated nitric acid. (This may take some time and should be provided if this is the method used.) A 0.100 g sample of manganese is dissolved in about 20 mL of 1:1 nitric acid. It is then diluted to 1 L. Appropriate aliquots are then pipetted into small beakers, treated with 5 mL of the special reagent, hydrogen peroxide, and ammonium persulfate, as with the samples, and finally diluted to volume (100 mL is convenient).

Literature Cited

1. *Toxicological Profile for Manganese*, U.S. Department of Health and Human Services, Public Health Service, Agency for Toxic Substances and Disease Registry, TP-91/19, Atlanta, GA, July 1992.

2. Susan Budavari, Editor, *The Merck Index*, 11th ed., Merck and Company, Inc., Rahway, NJ, 1989.

3. W. C. Cooper, The health implications of increased manganese in the environment resulting from the combustion of fuel additives: a review of the literature, *J. Toxicol. Environ. Health*, 14:23-46 (1984).

4. G. C. Cotzias, Manganese in health and disease, *Physiol. Rev.*, 38(3):503-533 (1958).

5. A. E. Greenberg, L. S. Clesceri, and A. D. Eaton, *Standard Methods for the Examination of Water and Wastewater,*3-75 to 3-76, American Public Health Association, Washington, DC, *1992.*

6. G. R. Holz, R. J. Huggett, and J. M. Hill, Behavior of manganese, iron, copper, zinc, cadmium and lead discharged from a wastewater treatment plant into an estuarine environment, *Water Res.*, 9:631-636 (1975).

7. A. Kabata-Pendias and H. Pendias, *Trace Elements in Soils and Plants,* CRC Press, Boca Raton, FL, 1984.

8. A. J. Paulson, R. A. Feely, H. C. Curl, et al., Behavior of iron, manganese, copper and cadmium in the Duwamish River estuary downstream of a sewage treatment plant, *Water Res.*, 18:633-641(1989).

9. E. J. Underwood, Manganese, In: *Trace Elements in Human and Animal Nutrition,* 3rd ed., Academic Press, New York, 1971, pp. 177-203.

Experiment 10

Determination of Trace Amounts of Metals by Atomic Absorption Spectrometry

Objective—The objective of this experiment is to introduce one of the most important methods of environmental chemistry, atomic absorption spectrometry. This instrumental method is used to detect metals and metalloids down to the ppb level. The method is fast and accurate and can be made to be essentially free of interferences.

Introduction—Free atoms cannot undergo rotational or vibrational energy transitions, as molecules can. Only electronic transitions can occur when energy is absorbed or emitted. Because electronic transitions are discrete (quantized), line spectra are observed.

There are various ways of obtaining free atoms and measuring the radiation they absorb or emit. In flame spectrometry, a solution is aspirated into a flame and the compounds present are thermally dissociated into atomic vapor. The heat of the flame first causes the solvent to evaporate. The micro-crystals produced are partially (or wholly) dissociated into the elements in the gaseous state. Some of the atoms thus produced can absorb radiant energy of a particular wavelength and become excited to a higher electronic state. When these atoms fall back to lower energy levels and emit light, this provides the basis for a very sensitive analytical technique, **atomic emission spectroscopy**. In this method high-temperature electric arcs or plasmas are used to maximize the production of excited atoms.

The term "**atomic absorption**" refers to the absorption of energy from a light source, with a consequent decrease in the radiant power transmitted through the flame. The measurement of this absorption corresponds to **atomic absorption spectrometry.**

The majority of atoms in a flame are in the ground state; thus, most electronic transitions originate from this state. A partial energy-level diagram for lithium is shown in Figure 10-1. There are several possible transitions for lithium, but the primary line is at 670.890 nm.

More than 60 elements can be determined by atomic absorption spectrometry, many at the part per billion level. Only metals and metalloids can be determined directly by usual flame methods because the resonance lines for nonmetals occur in the vacuum ultraviolet region of the spectrum. Table 10-1 lists the atomic absorption detection limits and wavelengths used for several environmentally important elements. For analytical measurements, concentrations should be at least 10 times the detection limit since, by definition, the precision at the detection limit is no better than ±50%.

Figure 10-1 Partial Energy Level Diagram for the Lithium Atom

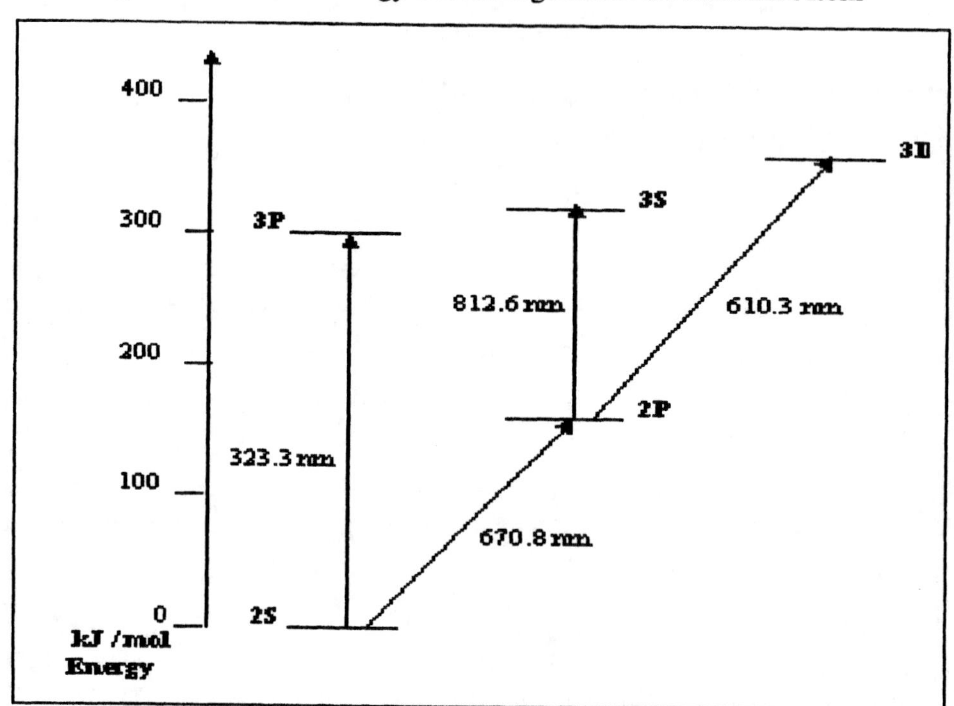

Instrumentation

Atomic absorption uses essentially monochromatic radiation to excite vaporized atoms in their ground state. The instrument consists of a light source, a cell (consisting of the aspirated sample), a monochromator, and a detection system. The instrument is shown in Figure 10-2.

The source, usually a hollow cathode tube, emits essentially line radiation of the same wavelength as that being absorbed by the element under study. This is accomplished by making the source out of the sample element. Thus, if iron is to be determined, a lamp having an iron cathode is used.

The sample is nebulized into a premixed gas-air burner designed for a long path length. The radiation then passes into a monochromator and is measured at the detector. The amount of radiation absorbed is proportional to the concentration of the element in the sample. A calibration curve is obtained by measuring the absorbance of a series of standard solutions, as was done in Experiment 9.

Table 10-1 Wavelengths and Detection Limits for Various Elements

Element	Wavelength (nm)	Detection Limit (ppb)*
Ag	328.07	1
Ca	422.67	2
Cd	228.80	1
Co	240.72	2
Cr	357.87	2
Cu	324.75	4
Fe	248.33	4
Hg	253.65	500
K	766.49	3
Mg	285.21	3
Mn	279.48	0.8
Na	589.00	0.8
Ni	232.00	5
Pb	283.31	10
Sn	235.48	50
Sr	470.73	5
Zn	213.86	1

*Values in the table refer to an air-acetylene flame.

Source: *Analytical Methods for Atomic Absorption Spectrophotometry*, Perkin Elmer Corporation, Norwalk, CT, 1982.

Experimental Methods

Atomic absorption spectrometry is used for the analysis of a great variety of substances, containing trace elements (ppb concentrations) as well as major (greater than 1%) inorganic constituents. Sample types include agricultural and biological samples, minerals, petroleum and its products, air, and water.

Some elements present in biological systems play an essential role when present in trace amounts. This is true for iron, zinc, boron, and manganese at levels of about 20 ppm, and for copper, cobalt, and molybdenum from 0.1 to about 5 ppm. Furthermore, some toxic elements can be bioaccumulated by animals and plants and include chromium, lead, mercury, arsenic, antimony, tellurium, and thallium.

Plants and other biological systems have an organic matrix. The digestion of these media can be accomplished by dry ashing at 450–500°C, followed by acid redissolution with 0.1 M HNO_3. Ashing methods run the risk of losing volatile elements, such as arsenic, antimony, selenium, tellurium, cadmium, mercury, and lead. In this case, a wet digestion with oxidizing acids is used. Wet acid digestion can be carried out using a combination of concentrated nitric and sulfuric acids.

Biological fluids (blood or urine) can usually be aspirated after a simple dilution. In a blood analysis, serum or plasma is generally used since this fraction contains significant amounts of metals. However, an analysis to detect lead poisoning uses red blood cells since they concentrate lead. The metals usually measured in blood include potassium, zinc, iron, and magnesium.

Atomic absorption is the method of choice for water analysis. Water types frequently analyzed include spring and river waters, seawater, wastewaters, and industrial effluents. Elements frequently determined include cadmium, chromium, copper, arsenic, lead, mercury, selenium, and zinc.

Some precautions are necessary to obtain good results. Water samples require good filtration. In particular, river water can contain a colloidal suspension of clay and silica, the chemical constituents of which can interfere with the elements being measured. Also, if the anions in the sample are not known, the water should be acidified with an excess of 1% acid (such as HNO_3) to destroy carbonates and to create a uniform anionic matrix.

Glassware used to prepare samples for atomic absorption analysis should be thoroughly washed, after which it should be rinsed with concentrated, high-purity nitric acid. This is followed by rinsing with DI water and, if possible, nanopure water. Standard solutions should not be stored in glass containers because leaching will cause changes in the metal concentrations over time. Storing in plastic containers is recommended if the samples and standards are not going to be analyzed during the same lab period.

The sensitivity of atomic absorption is not always sufficient to measure concentration levels of ppb or less. In this case, separation or enrichment methods are used. By extracting organic complexes of the metal with a suitable solvent, it is possible to concentrate the trace element in a solution by a factor of 10–50. The extract can be analyzed directly.

Figure 10-2 A Basic Atomic Absorption Spectrometer

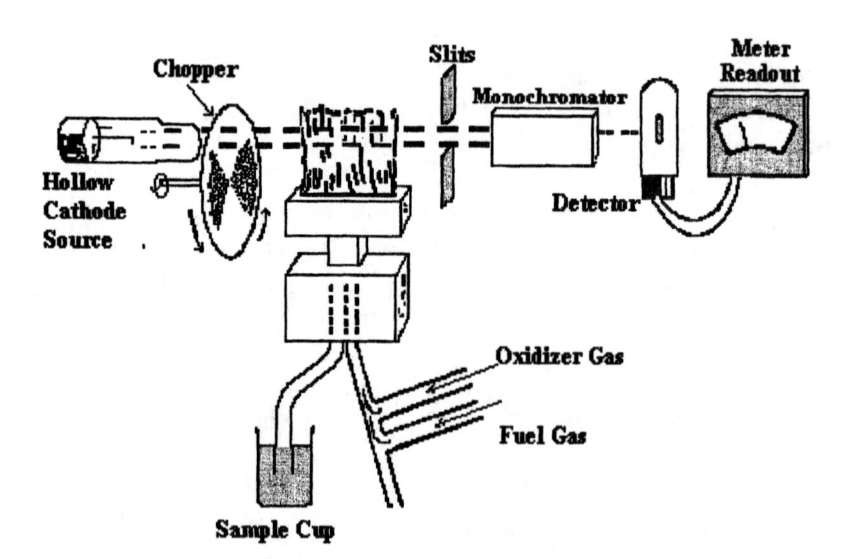

Source: G. H. Schenk, R. B. Hahn, and A. V. Hartkopf, *Introduction to Analytical Chemistry*, 2nd ed., Allyn and Bacon, Boston, 1981, p. 360. By permission.

Using the Atomic Absorption Spectrometer

The instrument consists of five main components, most of which are under microprocessor control on modern instruments. These are shown in Figure 10-2.

1. The hollow-cathode lamp power supply provides the controls for adjusting the current in the lamp. Each lamp has a maximum operating amperage which is indicated on the lamp. This current must not be exceeded or a very expensive lamp will be ruined.

2. The gas regulator panel controls and measures the flows of fuel and support (oxidant) gases. Each element requires a specific ratio of fuel to support gases, and specific conditions for each element are given in the methods manual for the instrument.

3. The burner assembly consists of the burner itself, the elevator for changing the burner height, and the aspiration system used to introduce the sample into the flame. The burner height is very important for each element.

4. The monochromator consists of the wavelength adjuster, a variable slit width, and controls used for zeroing the instrument and signal amplification.

5. The readout module provides a digital readout, a printer, or both.

Safety Issues

1. When heated sufficiently, nitric acid produces fumes of nitrogen dioxide. These attractive red-brown fumes are exceptionally toxic; 100 ppm levels are considered dangerous for even a short exposure and 200 ppm may be fatal. (See *The Merck Index*, [3], for further information on nitrogen dioxide.)

2. Safety glasses must be worn at all times in the chemistry laboratory.

3. When using the atomic absorption spectrometer, an instructor must be present.

4. Since air/acetylene mixtures can be explosive in certain ratios, it is essential that there be sufficient air and acetylene in the cylinders. Also, do not exceed the red line pressure on the reduction gauge of the acetylene regulator.

5. Before using the instrument, obtain instructions for shutting the instrument down quickly in the event of an emergency. Most modern instruments have built-in safety features, including a button or switch that shuts off the flow of gases immediately.

Locate the shut-off valve on the acetylene cylinder. This should be closed immediately in case of a fire.

6. The hood above the instrument must be on when aspirating samples. Otherwise vapors of metals and acid will be inhaled.

7. Biological samples, including foods, fish, plants, and even soils and sediments can be decomposed by dry ashing (followed by acid treatment, including perchloric acid) or by wet digestion with acids, including perchloric acid. However, it is highly recommended that any procedure calling for the use of perchloric acid **not be used!!** When dry, the resulting residue can be highly explosive. The authors have found that most samples can be digested, at least to the extent of freeing the desired metals, with concentrated nitric acid or with a mixture of nitric and sulfuric acids, and extensive heating.

8. Concentrated nitric acid is strongly oxidizing and should be kept away from organic matter (except for the sample, of course). In contact with some organic matter, it can be explosive. It is corrosive to the skin, and all contact with it should be avoided. Also, inhaling its fumes can cause severe respiratory distress. Fumes from the heating of nitric acid contain nitrogen dioxide, which is exceptionally toxic and is discussed in Safety Precaution Number 1.

9. Concentrated sulfuric acid at room temperature is not oxidizing. However, it is exceedingly hygroscopic, absorbing rapidly any water it comes in contact with. On the skin it rapidly absorbs water and can cause permanent damage in seconds. Use the utmost care in handling it!

10. The metals from the environment that are frequently analyzed include mercury, lead, cadmium, and many other heavy metals that are extremely toxic. In making up standard solutions of these and other toxic materials, use gloves and work on paper towels or other protective material so that if there is spillage, it will not get on the lab desk top. Most of these metals can cause serious problems even at the low part per million level, or even lower in the case of mercury.

11. Make sure that there is a loop in the drain tube and follow the instructions in the instrument manual on whether the drain tube in the waste vessel should be submerged or not.

Procedure—This experiment is intended to illustrate the basic procedures used to analyze real samples for metals at the low part per million level. Metals found in quantity in river sediments include iron, manganese, lead, and strontium. Samples taken near shipyards are often rich in chromium, cadmium, lead, titanium, and other metals. Seashells also provide interesting samples as they accumulate significant amounts of lead, strontium, and other metals.

1. Prepare a sediment sample by first decanting off excess water.

2. Tare a labeled 150 mL beaker on an analytical balance and scoop in 9–11 g of wet sediment. Weigh the sample to the nearest 0.1 mg. Repeat for a different environmental sample. (Instead of using a wet sample, a sample can be dried (time-consuming) or a dry weight of the sample can be determined as described in Experiments 2 and 9. Also, the core segments prepared in Experiment 2 can be used in this experiment.)

3. In a fume hood, add 10 mL DI water and 10 mL of high-purity (for trace metal analysis) concentrated nitric acid to each sample. Add the acid slowly if there is frothing.

4. Prepare a blank by adding 10 mL of DI water and 10 mL of high-purity concentrated nitric acid to another 150 mL beaker. Treat the blank identically to the samples.

5. Cover the beakers with watch glasses and gently swirl to mix. Heat on a hot plate to just below the boiling point and continue heating for 30 minutes. If necessary, add an additional 10 mL of acid if it appears that organic matter has not decomposed. Also, add additional water if there is much evaporative loss. (As mentioned in Experiment 9, the NO_2 fumes given off by decomposing nitric acid are extremely toxic. Avoid breathing in any of these fumes. Refer to Safety Issues 1.)

6. Remove the beakers from the hot plate, add 10 mL DI water to each sample, and allow to cool for 5 min. Filter each sample, including the blank, using medium speed filter paper, catching the filtrate in a 100 mL volumetric flask (see Note 1). Rinse the beaker twice with small portions of DI water, adding the rinses to the funnel. Finally, rinse the funnel twice, using small amounts of DI water from a squeeze bottle. Bring the volume in the flask up to the mark, stopper, and mix thoroughly. The solution should be clear and devoid of any particulates. Transfer the samples and blank to plastic bottles if they are not going to be immediately analyzed.

7. Prepare four standard solutions having concentrations of 1, 2, 4, and 8 ppm of the metal under study. They are made by diluting a 100 ppm (0. 1 mg/mL) stock solution using pipets and volumeric flasks. Prepare 100 mL of each standard and transfer to plastic containers for storage if samples and standards are not going to be immediately analyzed. (**Note to the instructor:** It is convenient to have different groups make up standards of different metals and the entire class can use the complete sets of standards. That way each group can determine several metals in their sample.)

8. Familiarize yourself with the instrument by locating the various components, referring to Figure 10-2 and the instrument itself. The instructor should give an introductory presentation on the adjustments that must be made and on using the instrument's software.

9. Zero the instrument before making readings on samples or standards. Then measure the absorbance of the blank. If the blank does not read zero, aspirate more water and read the blank again.

10. Determine the absorbance of the standards and then the absorbance of the samples (see Note 2). If the absorbance of a sample is higher than the absorbance of the most concentrated standard, dilute the sample to bring it into the range of the standards. If standards from other groups are used, switch to the next lamp, and run the blank, followed by the standards and samples as before.

11. After samples have been measured, again measure the absorbances of some standards for each metal analyzed to check the stability of the instrument.

12. Aspirate DI water into the flame for 5 minutes after running your last sample to clean the burner assembly.

Waste Minimization and Disposal

1. Sediment or soil samples from previous experiments can be used. If new samples are collected, do not collect more than 100 g to minimize disposal efforts and costs.

2. Standard solutions of toxic metals must be poured into an appropriate waste receptacle, and the waste will be properly treated or disposed of (a commercial hazardous waste contractor, for example).

Data Analysis

1. Prepare a calibration plot of A (ordinate) versus concentration of the metal for the standards (if the instrument's computer does not generate one). This assumes a zero absorbance for the blank. If this is not the case, the net absorbance is used. Repeat for each metal tested.

2. If the instrument's computer does not calculate the ppm of the analyte, use the calibration plots to calculate ppm of the metal. Repeat for each sample and then repeat for the other metals. Alternatively, use computer software to carry out a least-squares analysis of the data for the standards. The results of the analysis can then be used to write an equation for the straight line, and this in turn can be used to calculate the sample solution concentrations. The product of this concentration times the volume of the solution gives the total amount of extracted metal (see Step 3).

3. Calculate the (mg of metal)/(kg of sediment) for each metal and for each sample. The amount of analyte, X, in the sediment (soil) is obtained by first multiplying the measured extract concentration (mg/L) of X times the volume, V of the extract solution, in liters:

$$\text{mg}\,X = (\text{mg/L}) \times V \tag{10-1}$$

Then the mg of X per kg of sample is

$$(\text{mg/L})\,(V)\,/\,\text{kg sample} = \text{ppm}\,X \tag{10-2}$$

4. Discuss the significance of your results. Do your values fall within expected values, or are they much higher than normal background levels? Remember to cite your sources for "normal" levels.

Supplemental Activity

1. If time allows, the student can observe the effect of changing the following parameters by aspirating a concentrated standard: changing the lamp alignment, changing the burner height, increasing the flow of acetylene, increasing the flow of air.

2. After aspirating a concentrated sample of a metal and determining its absorbance, add 1–3 drops of 85% phosphoric acid to about 10 mL of the sample and reaspirate it to observe the effect of complexation.

3. Because of similar size, Pb^{2+} is often found associated with Ca^{2+}. Elevated lead levels have been found in calcium-based antacids and calcium supplements. A possible small project is the determination of the lead content of those products. The authors have also found strontium to be at substantial levels in these products.

Questions and Further Thoughts

1. In dilute aqueous solution, why is the numerical value of ppm the same as mg/L?

2. Why do sediment samples taken near shipyards usually have elevated levels of many metals, especially lead, cadmium, and chromium?

Notes

1. Particulates must be scrupulously eliminated from samples before aspirating them into the burner. Otherwise, the aspirator may become plugged. Medium porosity filter paper is sufficient for this purpose (Whatman Number 1, for example).

2. If absorbances fluctuate, the most likely cause is poor aspiration due to a clog. See the instructor about this problem. The use of a wire supplied with the instrument can be used to dislodge the obstacle.

3. There must be a good, fully functional hood over the flame. **The hood can get very hot!**

4. Some metals should not be studied at all by beginning students or even advanced students. One such metal is beryllium, which can cause serious lung disease even at extremely low concentrations.

5. The final results of this experiment are mg/kg wet sediment. Many literature results are given in this form, but mg/kg dry sediment is preferable since the water content of samples can be quite variable. The dry weight basis can be determined by either:

   a) Drying the sample before digesting. This can be lengthy.

   b) Taking a 1 g sample of wet sediment and drying to constant weight at 110°C. This takes less than an hour. Letting the ratio, (dry weight)/(wet weight) = *f*, the dry weight equivalent of a sample is dry weight = (*f*)(sample wet weight), and then,

$$\textbf{mg/kg dry weight} = \textbf{(mg/L)}\ V\textbf{/kg dry sample} \tag{10-3}$$

Literature Cited

1. *Analytical Methods for Atomic Absorption Spectrophotometry,* Perkin Elmer Corporation, Norwalk, CT, 1982.

2. B. J. Alloway and D. C. Ayres, *Chemical Principles of Environmental Pollution,* Blackie Academic and Professional, New York, 1993.

3. S. Budavari, Ed., *The Merck Index,* 11th ed., Merck and Company, Inc., Rahway, NJ, 1989.

4. G. W. Ewing, *Instrumental Methods of Analysis,* 5th *ed.,* McGraw-Hill, New York, 1985.

5. J. S Fritz and G. H. Schenk, *Quantitative Analytical Chemistry*, 5th ed., Allyn and Bacon, Boston, 1987.

Experiment 11

Removal of Chromium(VI) From Wastewater at the Part Per Million Level

Objectives—The objectives of this experiment are:

1. To illustrate how harmful wastes can be converted into harmless materials

2. To illustrate the method of waste reduction

Introduction—The technology for the treatment and disposal of hazardous wastes is the most rapidly developing area of environmental engineering. If possible, chemical wastes are treated on site, using chemical reactions to transform pollutants into less harmful substances. The full range of chemical reaction types—neutralization, precipitation, redox, and complexation—is used. This experiment will illustrate the principles used to eliminate an important pollutant, chromium(VI).

Heavy metals can frequently be detoxified by precipitation. General solubility rules indicate that many heavy metal salts are insoluble. In particular, heavy metal sulfides, sulfates, and carbonates as well as the hydroxides are very insoluble in water. Lead sulfide, for example, has a K_{sp} of 8.0×10^{-29} at 25°C, and chromium(III) hydroxide has a K_{sp} of 6.3×10^{-31} at 25°C.

In this experiment a common and important industrial pollutant, chromium(VI), will be reduced to chromium(III), followed by precipitating the hydroxide. Chromium is one of a number of elements (including aluminum, lead, and zinc) that is amphoteric and forms a complex with excess hydroxide to yield a soluble species. Thus, in excess base chromium(III) hydroxide acts as a Lewis acid:

$$\mathbf{Cr(OH)_3(s) + OH^-(aq) \rightleftharpoons Cr(OH)_4^-(aq)}$$

Theory The solubility of $Cr(OH)_3$ in water can be calculated by using the K_{sp} expression. Letting S be the molar solubility,

$$K_{sp} = [Cr^{3+}][OH^-]^3 = 6.3 \times 10^{-31} = (S)(3S)^3 \qquad (11\text{-}1)$$

If it is assumed that water does not contribute to the concentration of hydroxide, then $S = 1.2 \times 10^{-8}$ mol/L (or 0.64 ppb of Cr^{3+}). The hydroxide from the precipitate will have a concentration of $3S$, or 3.7×10^{-8} mol/L, whereas pure water has a hydroxide concentration of 1.0×10^{-7} mol/L. Therefore, the concentration of chromium from the trivalent hydroxide would be less in pure water than the value calculated above; thus, the exact value must be calculated in conjunction with the K_w of water.

As the concentration of OH^- is increased, the Cr(III) concentration decreases (the common ion effect). However, at a sufficiently high pH, $Cr(OH)_4^-$ begins to form. The formation of this complex is given by the equation

$$Cr^{3+}(aq) + 4\,OH^-(aq) \longrightarrow Cr(OH)_4^-(aq)$$

and the equilibrium constant is written

$$K_f = [Cr(OH)_4^-] / [Cr^{3+}][OH^-]^4 \qquad (11\text{-}2)$$

The equilibrium constant, K_f, is called a formation constant and for the above complex is equal to 8.0×10^{29} at 25°C. The large value indicates that the complex is a very stable species.

Complexation is very important in environmental waters as both natural complexing agents (chlorides, phosphates, tannins) and pollutants (phosphates, cyanide, and nitrilotriacetic acid, NTA) can cause normally insoluble metal precipitates to become reentrained into the water column.

Chromium(VI) can be reduced to chromium(III) in acid solution by many reducing agents. Four of the more important ones are sulfur dioxide, ferrous sulfate, sodium bisulfite, and sodium metabisulfite. With sulfur dioxide, the reaction proceeds as shown:

$$SO_2(g) + H_2O(l) \longrightarrow H_2SO_3(aq)$$

and

$$3H_2SO_3(aq) + H_2Cr_2O_7(aq) \longrightarrow Cr_2(SO_4)_3(aq) + 4H_2O(l)$$

Following reduction, $Cr(OH)_3$ is precipitated with base:

$$Cr^{3+}(aq) + 3OH^-(aq) \longrightarrow Cr(OH)_3(s)$$

The chromium(III) hydroxide takes up little space compared with the original solution, and it can be kept in a labeled container until it is ultimately disposed of. It is not considered to be hazardous waste.

Properties of Chromium

Chromium is a widely distributed element found in rocks, animals, plants, and soil. Chromium metal is commercially produced by the reduction of chromite ore using carbon, aluminum, or silicon. Chromite ore is imported, mainly from South Africa, Zimbabwe, and Turkey.

Metallic chromium is mainly used for making steel and other alloys and compounds of the element. Chromium salts are widely used in industrial processes, including the manufacture of dyes and tanning leather and as a wood preservative. Chromates are often added to water for corrosion resistance, as in automotive and marine diesel cooling systems and in cooling tower water. Chromium is used in paint pigments and in protective metal coatings.

Chromium in the Environment

Chromium enters the atmosphere from burning coal and oil, steel production, and chromium manufacturing. In air, chromium compounds are mainly associated with fine dust particles that eventually settle out after remaining in the atmosphere for up to 10 days. The level of chromium in air is generally low, with concentrations of 10–30 ng/m^3.

Although most chromium waste from plating operations is recovered, large amounts of chromium-containing wastewaters have been discharged into surface waters. At levels greater than 3 ppb in water, industrial sources are indicated. A large amount of chromium enters sewage treatment plants from industrial and residential sources.

Chromium is normally found as chromium(VI) in polluted waters. Chromium(VI) in water will eventually be reduced to chromium(III) by organic or other reducing matter in the water. When it is found in natural waters, chromium(III) exists mainly as $Cr(OH)_3$. Trivalent chromium is rarely found in potable water, and when it is, its concentration is often less than 2 ppb. Although most of the chromium in water eventually settles into the sediment, a small amount remains in the water column and can persist for years.

Chromium(III) in soil is present mainly as the insoluble carbonate and oxide; thus, it will be immobile. However, its solubility and its mobility may increase due to the formation of soluble complexes with organic matter in the soil.

High levels of chromium in soil are due to the disposal of chromium-containing products, wastes from factories, and coal ash from electric utilities. Landfilling has been the most important method for the disposal of chromium wastes generated by industry. In the chemical industry chromium wastes occur predominantly as chromium(VI). Pretreated sludge containing chromium(III) is disposed of by land burial. Ocean dumping was used in the past but now is illegal!

All plant and animal tissues appear to contain chromium. Typically, vegetable tissues contain 10–500 ppb chromium. Fish do not significantly accumulate chromium from water. Levels in oysters, mussels, clams, and other mollusks range between 0.1–6.8 mg/kg (dry weight). Fish and shellfish collected from ocean dumping sites off the United States east coast contained 0.3–2.7 mg/kg (wet weight).

Chromium as an Essential Element

In the body, chromium(III) is able to form complexes with nucleic acids and proteins, but chromium(VI) does not do so. Without chromium in the diet, the body loses its ability to use sugars, proteins, and fats properly, resulting in weight loss, decreased growth, improper function of the nervous system, and a diabetic-like condition. In humans and other animals, chromium(III) is thus an essential nutrient. It is the active ingredient of the glucose tolerance factor and functions by facilitating interaction of insulin with receptor sites, influencing metabolism.

Effects of Chromium on Humans

For humans, food is the greatest source of chromium. Chromium(III) occurs naturally in many vegetables, fruits, meat, and grains. Acidic foods in contact with stainless steel cans or cooking utensils may contain increased levels of chromium due to leaching. On average, adults in the United States take in about 60 μg of chromium daily, and the officially recommended level for adults is 50–200 μg/day. The EPA has set the maximum level of total chromium allowed in drinking water at 50 μg/L.

High levels of chromium occur in the following areas:

1. Landfill sites containing chromium-containing wastes

2. Waterways with outfalls from industrial sources including electroplating, leather tanning, and/or textile industries

3. Roadways, since automotive emissions contain chromium from brake lining and catalytic converters

4. Smoking areas, since tobacco contains 0.24–14.6 mg/kg chromium

At elevated levels, chromium may cause harmful health effects. Chromium(VI) is an oxidizing agent and a carcinogen in humans and experimental animals. It enters the body more easily than chromium(III) but, once inside the body, is converted to chromium(III). Workers handling substances containing chromium(VI) have developed skin ulcers. However, no studies have shown that chromium compounds cause cancer after dermal or oral exposure. Most ingested chromium leaves the body within a few days through the feces and never enters the bloodstream.

Inhaled chromium will enter the bloodstream, pass through the kidneys, and be eliminated in the urine within a few days. Long-term exposure to chromium is associated with lung cancer in workers exposed to high levels in air. Lung cancer may occur years after exposure to chromium. Occupational exposures to chromium compounds result in increased risk of respiratory system cancers, notably bronchogenic and nasal. In the lungs, ascorbate reduces chromium(VI) to chromium(III), resulting in a decreased residence time of chromium in the lungs. This constitutes a first defense against oxidizing agents in the lungs.

The first defense against chromium(VI) after oral exposure is reduction to chromium(III) in the gastric environment, where ascorbate again plays an important role. Chromium(VI) has been shown to cause severe liver effects in workers in the chrome-plating industry. The spleen has been shown to concentrate chromium with time. In the human body, chromium has been detected in the hair, urine, serum, red blood cells, and whole blood.

Safety Issues

1. Safety glasses must be worn at all times when working in the laboratory or in the field.

2. When handling chromium-containing solutions, wear protective gloves.

3. Refer to the safety precautions in Experiment 10 for atomic absorption analysis.

Procedure—A composite waste solution containing chromium(VI) at the ppm level will be provided. Although several reducing agents are capable of reducing chromium(VI) to chromium(III), sodium metabisulfite will be used in this experiment:

$$3Na_2S_2O_5\,(aq) + 2H_2Cr_2O_7\,(aq) + 3H_2SO_4\,(aq) \longrightarrow 3Na_2SO_4(aq) + 2Cr_2(SO_4)_3(aq) + 5H_2O\,(l)$$

Following reduction, Cr(III) is precipitated with $Mg(OH)_2$,

$$Cr_2(SO_4)_3(aq) + 3Mg(OH)_2(s) \longrightarrow 2Cr(OH)_3(s) + 3MgSO_4(aq)$$

Sodium hydroxide could be used for the precipitation but yields a gelatinous precipitate which, is difficult to filter. Although a good precipitate can be obtained by proper regulation of the pH, $Mg(OH)_2$ automatically produces the correct pH for a good precipitate to form. The clear filtrate will be slightly basic (pH 7.1–9.2) and contains less than 0.25 ppm of chromium(VI) and only trace amounts of chromium(III).

Disposal of Solutions of Chromium(VI) Waste

Chromium(VI) wastes are very common in industry and in university chemistry labs. The usual disposal procedure in the past was to flush such wastes down the drain, and if the chemist was conscientious, he or she would wash it down with a large amount of water. It is now incumbent on the instructor and the institution's safety officer to check with local, state, and federal agencies to verify what substances, if any, can be flushed down the laboratory drain. The best practice is for very toxic substances to be detoxified before disposal.

1. In this experiment, you will be given 500 mL of a solution containing chromium(VI) at a concentration between 200 and 1000 ppm. Pipette exactly 1 mL of your unknown into a 100 mL volumetric flask and dilute to the mark with DI water. Your solution is now between 2 and 10 ppm in chromium.

2. Four solutions of chromium in the same form as your unknown (chromate or dichromate) having concentrations between 2.5 and 15 ppm will be provided for this experiment. Using the standard solutions, estimate the concentration of chromium in your diluted unknown by comparing the intensity of the color of your solution with the standards. Verify the concentration of your diluted unknown by using atomic absorption analysis, using the four standards to construct a calibration plot. Save samples and standards, as they will be needed again in Step 8.

3. Measure 200 mL of your original, undiluted solution into a beaker using a graduated cylinder. Add a magnetic stir bar, and 3 mL of 2 M sulfuric acid and stir on a magnetic stir plate. Note any color change upon acidification. Estimate the pH using test paper. If not 4 or less, add additional 1 mL aliquots of sulfuric acid to bring the pH to the desired value.

4. Add 15 mL of 3.5% sodium metabisulfite in water to the unknown in the beaker. Stir for 0.5 hour, and frequently note the color of the stirred mixture. Check the pH again and if it is not between 4 and 5, add 1 mL more of the sulfuric acid. If at this point a yellow tinge can be discerned, add 1 mL of the metabisulfite solution to eliminate the color.

5. Add 5 g of magnesium hydroxide to the beaker and continue to stir. Note any color change. Chromium(III) hydroxide is yellow-brown in color, and if a green color persists, this is due to some remaining chromium(VI).

Check the pH to be certain that it is 8 or higher. If not, add additional magnesium hydroxide until the proper pH is attained. After achieving the correct pH, stir for 5 more minutes and then allow the precipitate to settle for a few minutes by turning off the stirrer.

6. Test the supernatant liquid for the presence of chromium(VI) by pouring a few mL into a small beaker and adding a few drops of acidified 0.1 g/L potassium iodide solution. When a few drops of starch solution are added, a blue color will appear if there is any Cr(VI) left. If there is, the mixture must again be brought to the correct acidic pH followed by the addition of more metabisulfite and, finally, additional magnesium hydroxide.

7. Use a Büchner funnel and a clean filter flask to suction filter the excess magnesium hydroxide and the chromium(III) hydroxide. Note the color of the precipitate. The filtrate should be clear. If not, add a few drops of 2 M sulfuric acid, followed by just enough sodium meta-bisulfite to make the solution clear. Follow with a few tenths of a gram of magnesium hydroxide and refilter the mixture. Again, the filtrate should be clear.

8. Determine the concentration of chromium in the filtrate by atomic absorption using the standards from Step 2. Both atomic absorption analyses can be done at the same time so that repeat measurements of the standards will not be necessary.

Waste Minimization and Disposal—Any chromium(VI) waste should be given to the instructor for disposal. The chromium hydroxide generated in this experiment is not considered to be a hazardous waste. However, as it takes up little space, it can be collected for a long time and not be a problem. It must at least be labeled properly.

Data Analysis

1. a. Report the ppm chromium in the diluted unknown as estimated by visual comparison with the standard solutions.

   b. Prepare a plot of absorbance (as ordinate) vs. ppm for the standards from the atomic absorption measurements.

   c. Use the previous plot to estimate the ppm of chromium in the diluted unknown and compare with the estimated value from the visual comparison of the unknown with the standards in part (a).

   d. Use your result from part (c) to calculate the ppm of chromium in your original, undiluted solution.

2. Use the calibration plot of part (1b) to calculate the ppm chromium in the filtrate remaining after precipitating out $Cr(OH)_3$.

Supplemental Activity

1. It is instructive to have the student estimate the lowest detectable concentration of chromium(VI) he or she can detect with the eye by standard comparison and use the instrument manual to find the minimum detectable limit for the instrument (IDL). It is noteworthy that accuracy and precision near the instrument detection limit are extremely poor, since this value is the least amount of analyte the instrument can differentiate above the noise level.

2. The instrument detection limit (IDL) should not be confused with the method detection limit (MDL). The IDL is the minimum amount of analyte the instrument can see under the experimental conditions. The MDL also takes into consideration all aspects of the method including extraction or digestion, any clean-up procedures, and the type and conditions of analysis.

Questions and Further Thoughts

1. Why do metals in high oxidation states enter the body more readily than metals in lower oxidation states?

2. By combining the K_{sp} for chromium(III) hydroxide and the K_w for water at 25°C, calculate the solubility of the hydroxide in pure water.

3. Can you calculate the pH at which the hydroxide precipitate dissolves to form the tetrahydroxochromate(III) ion? How could you verify this experimentally?

Notes—Atomic absorption analysis can be used to experimentally measure K_{sp} values. However, the solubility of the compound must be such that the metal concentration be above the IDL. The student may be given the task of finding chromium compounds that have K_{sp}'s that could be measured by flame AA.

Literature Cited

1. K. Akatsuka and L. T. Fairhall, The toxicology of chromium, *J. Ind. Hyg., 16*:1–24 (1934).

2. R. A. Anderson, Nutritional role of chromium, *Sci. Total Environ., 17(1)*:13-29 (1981).

3. A. M. Baetjer, C. Damron, and V. Budacz, The distribution and retention of chromium in man and animals, *Am. Med. Assoc. Arch. Ind. Health, 20*:136-150 (1959).

4. Susan Budavari, Ed., *The Merck Index,* 11th ed., Merck and Company, Inc., Rahway, NJ, 1989.

5. L. M. Calder, Chromium contamination of groundwater, *Adv. in Environ. Sci. Technol., 20*:215–229 (1988).

6. L. Fishbein, Overview of analysis of carcinogenic and/or mutagenic metals in biological and environmental samples, I. Arsenic, beryllium, cadmium, chromium, and selenium, *Int. J. Environ. Anal. Chem.,17(2)*:113–170 (1984).

7. H. M. Freeman, Editor in Chief, *Standard Handbook of Hazardous Waste Treatment and Disposal*, McGraw-Hill, New York, 1988, pp. 7.21–7.25.

8. R. Fukai, Valency state of chromium in seawater, *Nature 213*:901 (1967).

9. J. T. Kumpulainen, W. R. Wolf, C. Veillon, and W. Mertz, Determination of chromium in selected United States diets, *J. Agric. Food Chem., 27(3)*:490–494 (1979).

10. T. H. Lim, T. Sargent, III, and N. Kusubov, Kinetics of trace element chromium(III) in the human body, Am. J. Physiol., 244(4):R445-R454 (1983).

11. T. Norseth, The carcinogenicity of chromium and its salts, *Brit. J. Ind. Med., 43*:649-651 (1986).

12. W. B. Qi and L. Z. Zhu, Spectrophotometric determination of chromium in wastewater and soil, *Talanta, 33(8)*:694-696 (1986).

13. *Toxicological Profile for Chromium,* U.S. Department of Health & Human Services, Public Health Service, Atlanta, GA, (1992).

14. U.S. EPA, National interim primary drinking water regulations, *Fed. Reg., 40* (248):59566–59587 (1975).

Experiment 12

Solubility of Oxygen in Pure and Natural Waters and Its Dependence on Temperature and Salinity

Objective—The objective of this experiment is to determine the solubility of oxygen in pure water at several temperatures and to use these data to calculate the standard enthalpy and entropy changes occurring when oxygen dissolves in water. The solubility of oxygen in solutions of varying salinity at one temperature will also be determined.

Introduction—The determination of dissolved oxygen (DO) is one of the most frequently made environmental measurements. In addition to its importance for aquatic life, it is an excellent indicator of water quality. The DO is never high, even under the best of conditions. For water saturated with oxygen at 20°C and 1.00 atm pressure, the solubility is 2.8×10^{-4} mol/L (9.0 mg/L or 9.0 ppm). Existence for most aquatic organisms is difficult if the DO is less than 5-6 ppm, and for fish life ceases altogether if the DO is less than about 3 ppm. More active species, such as trout, require more oxygen, 5-8 ppm. Carp, on the other hand, survive DO levels as little as 3 ppm.

The intimate involvement of oxygen in life processes, photosynthesis and respiration, has prompted scientists to measure its concentration in a variety of environments. In natural waters depleted oxygen is replenished by aeration and photosynthesis. In rapidly moving waters, aeration is the most important process, whereas in lakes and ponds, where the water is less agitated, photosynthesis is predominant. Oxygen production is high during daylight due to photosynthesis, whereas at night there is no production but, in fact, there is consumption (respiration).

Organic matter of natural or human origin will reduce the amount of dissolved oxygen through the following reaction, which applies to carbohydrates:

$$(CH_2O)_x + xO_2(aq) \longrightarrow xCO_2(aq) + xH_2O(l)$$

The stoichiometry of this equation indicates that 12 g of carbon requires 32 g of oxygen for complete reaction. Thus, even a small amount of organic matter in water can decrease the DO significantly, if not all the way to zero. In areas where the DO is already low, rain storms can wash organic matter into streams and ponds and essentially deplete the oxygen, killing off many fish.

The solubility of oxygen in water depends primarily on three variables—pressure, temperature, and the concentration of dissolved salts. The first factor, pressure, is quantitated through Henry's law, which holds well over a range of pressures. Because dissolving oxygen in water is an exothermic process, there is a decrease in oxygen solubility with an increase in temperature. This behavior is true for most gases dissolved in water (Le Châtelier's principle).

The solubility of oxygen decreases with an increase in salt (electrolyte) concentration. This behavior is characteristic of many nonelectrolytes when an electrolyte is present in the solution and is known as the **salting-out effect.** The law relating the solubility of a gas and electrolyte concentration is

$$\ln (S_w / S_e) = K_s C_e \qquad (12\text{-}1)$$

where C_e is the electrolyte concentration, S_w is the solubility in pure water, S_e is the solubility in the electrolyte solution, and K_s is called the **salting coefficient**.

Theory—Partitioning is a very general phenomenon in chemistry and is one of its most important concepts. It is directly related to equilibrium, and thus to thermodynamics, and also to rate processes, including kinetics, diffusion, viscosity, etc. It is also the fundamental basis of all chromatographic methods and solvent extraction procedures.

In this experiment we examine the partitioning of molecular oxygen between the gaseous and liquid phases of a system. Although the empirical law governing this behavior (Henry's law) has been known for a long time, it is only one example of a more general law.

Suppose that a system can exist in two states, R and P, as shown in the equilibrium expression,

$$R \rightleftharpoons P$$

R and P represent the two possible states of a system (R and P are used to represent reactant side and product side). The system can be any type of equilibrium system.

The Boltzmann distribution law governs the way the system is distributed among the possible states. For two states, R and P, this law becomes,

$$N_P / N_R = \exp(-\Delta\varepsilon / k_B T) \qquad (12\text{-}2)$$

where N_P is the number of systems in state P, N_R is the number in state R, $\Delta\varepsilon$ is the energy difference, per particle, between the two states and T is the Kelvin temperature. The constant, k_B, is the Boltzmann constant. If $\Delta\varepsilon$ and k_B are both multiplied by Avogadro's number, N_A, the last expression becomes,

$$N_P / N_R = \exp(-\Delta E / RT) \qquad (12\text{-}3)$$

where $\Delta E = N_A \cdot \Delta\varepsilon$ and R is the gas constant.

The energy difference needs to be carefully defined since it is a different type of energy depending on the type of system and process involved. In this experiment, and indeed, for all chemical reactions and physical equilibria, the energy difference is the free energy difference, ΔG^o, where the superscript represents a standard reference pressure of 1 atm. Thus,

$$N_p / N_R = \exp(-\Delta G^o / RT) \qquad (12\text{-}4)$$

The free energy function is made up of two other thermodynamic functions, the enthalpy, ΔH^o, and the entropy, ΔS^o. The relationship is given by the Gibbs-Helmholtz equation,

$$\Delta G^o = \Delta H^o - T\Delta S^o \qquad (12\text{-}5)$$

The enthalpy change is the heat transferred between system and surroundings at a constant pressure. The entropy is related to the change in disorder of the system. The enthalpy and entropy changes are the two driving forces for changes to occur.

Equations 12-4 and 12-5 can be combined to give

$$N_P / N_R = [\exp(\Delta S^o / R)][\exp(-\Delta H^o / RT)] \qquad (12\text{-}6)$$

Dividing the number of particles by Avogadro's number and the volume gives molar concentrations, designated by brackets:

$$N_p / N_R = (N_p / N_A) / (N_R / N_A) = \text{mol}_p / \text{mol}_R = (\text{mol}_P / \text{V})/(\text{mol}_R / \text{V})$$

$$= [\text{P}] / [\text{R}]$$

The last ratio is simply the usual equilibrium constant, K_{eq}. Thus

$$K_{eq} = [\exp(\Delta S^o / R)][\exp(-\Delta H^o / RT)] \qquad (12\text{-}7)$$

Taking the natural logarithm of both sides gives,

$$\ln (K_{eq}) = -(\Delta H^o / R)(1 / T) + \Delta S^o / R \qquad (12\text{-}8)$$

This equation applies to any equilibrium, from the dissociation of weak acids and bases to the equilibrium vapor pressure of solids and liquids.

For a gas dissolved in water, the equilibrium expression is,

$$A(\text{g}) \rightleftharpoons A(\text{aq})$$

and equilibrium constant is,

$$K_{eq} = [A(\text{aq})] / [A(\text{g})] \qquad (12\text{-}9)$$

For a gas, it is more convenient to express its concentration in terms of partial pressure. Using the ideal gas law, a component of a gaseous mixture, *i*, has a partial pressure

$$P_i = n_iRT / V = [i]RT \qquad (12\text{-}10)$$

where [i] is the concentration of i in mol/L. Therefore, Eq. 12-9 can be written

$$K_{eq} = [A(aq)] / (P_A/RT) \qquad (12\text{-}11a)$$

or

$$[A(aq)] = (K_{eq}/RT)P_A = K_H P_A \qquad (12\text{-}11b)$$

where K_H is called Henry's law constant. Equation 12-11b shows that the concentration of dissolved gas is proportional to its partial pressure, which is Henry's law.

Safety Issues—Safety glasses must be worn at all times in the chemistry laboratory.

Procedure—Oxygen solubility will be determined with an oxygen electrode. Equilibrium saturation with oxygen occurs within minutes, so temperature stability for a short time is all that is required.

Measurements of DO will be made at approximately the following temperatures: 0°, 15°, 25°, 35°, and 45°C. Deionized water will be used, and saturation with oxygen will be achieved with a fritted bubbler attached via rubber tubing to an aquarium pump. Constant temperatures will be achieved by using a relatively large amount of water (250-400 mL). For elevated temperatures, a hotplate will be used and carefully adjusted so that heat losses equal the heat gained. (Alternatively, a hot water bath can be used.) It is possible to achieve reasonable temperature control in this fashion. Ice water will be used to achieve temperatures of 0° and 15°C. The solubility measurements will be repeated at room temperature using four solutions with varying salinity.

The dissolved oxygen content of several samples of natural waters will also be measured and compared with the expected value for pure water. Also, the saturated values for the samples will be determined. Measurements are corrected for the barometric pressure and temperature.

A: Calibration and Determination of Saturation Level of DO at 25°, 35°, and 45°C

1. Calibrate the dissolved oxygen meter at room temperature following the manufacturer's instructions, making sure that all specified conditions are met (any materials needed will be provided).

2. Add 250–400 mL of DI water to a 600 mL beaker.

3. Put a magnetic stir bar in the beaker and place the beaker on a magnetic stir hot plate. Adjust the stirrer for good mixing.

4. Attach a fritted bubbler to an aquarium pump using rubber tubing. Place the bubbler in the beaker of water so that the fritted glass is completely submerged.

5. Place the DO probe with a temperature sensor (or a thermometer, if necessary) in the center of the beaker at about half the depth of the water.

6. Bubble air through the water until a saturated (constant) DO is attained. Record the DO and the temperature.

7. Adjust the hotplate gradually until the temperature of the water is constant at about 35° C.

8. When the temperature is constant, measure the DO after its reading remains constant. Record the DO and temperature.

9. Repeat steps 7 and 8 at 45°C.

B: Determination of DO at 0° and 15° C

1. Set up the equipment as in Part A, Steps 2–5.

2. Note the temperature, and add small amounts of ice until approximately 15°C is reached.

3. When the temperature and the DO reading are constant, record the temperature and the DO.

4. Again add ice until the temperature is stable at approximately 0°C.

5. When the DO is stable, record the temperature and the DO.

C: Determination of DO Concentration for Four Solutions of Different Salinitiy

1. Prepare sodium chloride solutions having the following concentrations: 10^{-4} M, 10^{-3} M, 10^{-2} M, and 10^{-1} M. These may be prepared by diluting an appropriate stock solution (the stock solution or the four standard solutions may be provided).

2. Add 250 mL of one of the standard sodium chloride solutions to a 600 mL beaker.

3. Repeat Part A, Steps 3–6.

4. Repeat Steps 1–3 for each of the remaining three sodium chloride solutions.

D: Determination of the DO Concentration and the Saturation DO of Natural Waters

1. Add 250 mL of water from a natural source to a 600 mL beaker. This must be a freshly collected sample. Measure the DO and record the temperature of the sample.

2. Repeat Steps 3–6 of Part A for this sample.

3. Repeat Steps 1–2 for each environmental sample to obtain the DO and the saturated value.

Waste Minimization and Disposal—The only possible hazardous wastes in this experiment are the environmental water samples. These may be samples used in previous experiments, and they can be used in subsequent experiments. Any remaining samples will be disposed of according to the directions of your instructor.

Data Analysis

1. Report the solubility (ppm) of oxygen in water for each DI water sample measured, taking into account the ambient barometric pressure.

2. Use your results from Part 1 and Eq. 12-11b to calculate Henry's law constant, K_H, at the temperatures studied. Compare with the known values.

3. Plot the oxygen solubility (as ordinate) versus temperature (°C) for the DI water samples. If your computer software is able, curve-fit the data to obtain an equation relating oxygen solubility to the Celsius temperature.

4. Using your data from Part 1 and Eq. 12-11a, calculate K_{eq} for each temperature. Then plot ln K_{eq} (as ordinate) against $1/T$, preferably using computer software.

5. Carry out a least-squares analysis for the plot in Step 4 to obtain the slope, *y*-intercept, and correlation coefficient for the best straight line. Use these results and Eq. 12-8 to calculate ΔH^o and ΔS^o for the solution process. (The best result for the entropy is obtained by using the slope to calculate ΔH^o and then the equation of the straight line to calculate ΔS^o. This avoids a long extrapolation to obtain the entropy.)

6. Plot ln (S_w / S_e) (as ordinate) versus C_e and carry out a least-squares analysis to obtain the slope, *y*-intercept, and correlation coefficient.

7. Use the previous plot to obtain the salting coefficient, K_s. Comment on the linearity of the plot. What is the environmental significance of K_s? How does it compare with known values?

8. Make a table listing the measured DO and the saturation DO values for the environmental samples. Also give the temperatures of the measurements.

9. Discuss the measured and saturation DO values for the environmental samples and indicate how the location of the sample sites affects these values.

Supplemental Activity

1. To illustrate how slow re-aeration is for still water, obtain a sample of boiled DI water and measure its DO and temperature. Without bubbling or stirring, record the DO and temperature at 5-minute intervals for 30 minutes. Then slowly bubble air through the water and measure the DO and temperature at 1-minute intervals.

2. Repeat the previous procedure after placing a few drops of oil (kerosene, etc.) on the surface of the water in a beaker.

Questions and Further Thoughts

1. Why is the dissolving of oxygen an exothermic process? (Consider the steps in the solution process.)

2. If a temperature measurement is incorrect by 1°C, what difference would it make in a calculated solubility according to Eq. 12-11.

3. Why are power plants detrimental to the environment (i.e., what is thermal pollution)?

4. How does the pH of water affect its DO?

Notes

1. If erratic or unreasonable DO readings are obtained, the membrane of the DO electrode should be replaced according to the manufacturer's instructions.

2. The Winkler method for determining dissolved oxygen was devised in 1888 and is still the basis for most titrimetric methods of analysis. It is based on the oxidation of manganese(II) to manganese(III) by molecular oxygen, followed by titration with iodine.

Literature Cited

1. Michael L. Hitchman, *Measurement of Dissolved Oxygen,* Wiley, New York, 1978.

2. Gilbert M. Masters, *Introduction to Environmental Engineering and Science*, Prentice-Hall, Englewood Cliffs, NJ, 1991.

3. W. J. Moore, *Physical Chemistry,* 4th ed., Prentice-Hall, Englewood Cliffs, NJ, 1972.

Experiment 13

Determination of the Chemical Oxygen Demand of Natural Waters and Wastewaters Using a Standard Method

Objective—The objective of this experiment is to determine the chemical oxygen demand (COD) of natural waters or wastewaters using a potassium dichromate digestion, followed by colorimetric quantitation, as described in Standard Method 5220D (see Literature Cited, Ref. 2).

Introduction—Several easily measured properties are important indicators of the quality of a system of natural waters. These properties include total suspended solids (TSS), total dissolved solids (TDS), hardness, color, salinity, turbidity, dissolved oxygen (DO), biochemical oxygen demand (BOD), carbonaceous biochemical oxygen demand (CBOD), and chemical oxygen demand (COD).

In Experiment 12, the dissolved oxygen (DO) in pure and natural waters was measured and the maximum value was found to be about 9 ppm at 20°C. While this is a relatively small concentration of oxygen, it is essential for the health and survival of fish and numerous other aquatic species. If the DO level becomes 5 ppm or less, stressed fish and aquatic organisms can be killed entirely if rain causes runoff of organic matter into the water system, depleting the little remaining oxygen.

Theory—Not only is the DO level itself important, but the ability of a body of water to reduce this parameter, due to organic matter present, is important. This capacity is quantitated through the concept of BOD, the biochemical oxygen demand. In this procedure the DO of the water sample is measured and the sample is then incubated at 25°C for 5 days. At this time, the DO is measured again. The decrease in the DO, in mg/L, is the BOD. Thus the BOD is the quantity of oxygen consumed in the oxidation of organic matter in 5 days.

A BOD test provides the closest measure of the processes actually occurring in the natural water system. However, a test that takes 5 days is not convenient. Also, if the BOD is very high, the DO after 5 days will be zero since there is not enough oxygen available to oxidize all the organic matter present. Using an aliquot of the sample and mixing with an oxygen-saturated portion of pure water can circumvent this problem, but it is time-consuming and uncertain.

Another disadvantage of the BOD test is the presence of unknown factors in the test sample, including, but not limited to, the origin, concentration, and toxicity of pollutants and the number and viability of active microorganisms present to effect the oxidation of all pollutants. In many cases, there is either an insufficient number of microorganisms for the oxidation, or the microorganisms may be killed or rendered inactive by the type and/or concentration of pollutants present. The BOD test compensates for these factors by preparing multiple sample concentrations and the inoculation of these samples with one or more concentrations of a microorganism seed whose activity is measured during the experiment. Obviously, the preparation of multiple dilutions, often of two variables, is both tedious and time consuming. At the end of the 5-day incubation period, the experimental data may still be unusable because none of the experimental combinations met the method data criteria for DO depletion and residual DO. Repeating the test requires a new sample (the holding time for BOD samples is only 48 hours), new dilutions, and another 5 days of incubation.

Generally the BOD is due to decaying organic matter from plants and animals. In urban areas there is a large contribution from stormwater runoff, especially in areas of rapid development. The median BOD for unpolluted water in the United States is 0.7 mg O_2 / L.

The BOD measures the total oxygen demand for oxidizing all pollutants present. Depending upon the types of pollutants and microorganisms present, some of the oxidation process may occur as a result of bacterial oxidation of nitrogen-containing compounds. Addition of a nitrification inhibitor to the BOD process allows the oxidation depletion measurement of only the carbon-containing species. This measurement is called the carbonaceous biochemical oxygen demand (CBOD).

An alternative procedure to the BOD test is called the chemical oxygen demand (COD). A COD test uses a strong oxidizing agent, potassium dichromate, together with a small amount of silver ion to serve as a catalyst and a small amount of mercuric ion. Mercuric ion complexes chloride which would otherwise interfere in the test by being oxidized to Cl_2. Many natural waters, as well as the effluent from sewage discharges, are high in chloride ion.

In a COD test results can be obtained in 2 hours or less. Also, the method is simple and inexpensive. One disadvantage of the method is that dichromate can oxidize materials that would not ordinarily be oxidized in nature. However, like the BOD test, the COD test will not oxidize a number of refractory molecules. These frequently include aromatic ring systems and straight-chain aliphatics. However, oxidation with dichromate is 95–100% complete for most organic substances.

When wastewater contains only readily oxidizable organic matter and is free from toxins, the results of a COD test provide a good estimate of the BOD. When dichromate is used to oxidize organic matter in aqueous solution, the reduction half-reaction is,

$$\mathbf{Cr_2O_7^{2-} + 14H^+ + 6e^- \longrightarrow 2Cr^{3+} + 7H_2O}$$

When oxygen is used for the same purpose, the half-reaction is

$$\mathbf{O_2 + 4H^+ + 4e^- \longrightarrow 2H_2O}$$

Thus, it takes 6/4 or one and one-half times as much oxygen to carry out an oxidation as it would using dichromate (on a molar basis). This is used to relate the COD and the BOD.

Safety Issues

1. Safety glasses must be worn at all times in the chemistry laboratory.

2. Use the digestion block only in a fume hood and do not exceed its maximum recommended temperature. Shield the block and keep the door of the hood closed during digestion.

3. The vials contain concentrated sulfuric acid, potassium chromate (a strong oxidizing agent and possible carcinogen), and a very toxic mercury salt. Use gloves when handling these vials, and do not pour the remains down the drain when the test is complete. Follow the instructor's directions for disposing of the waste.

Procedure—There are several ways the COD test can be carried out. The student can prepare the chemicals, add the water samples, and reflux the mixture. The excess dichromate can be measured by titration or by a colorimetric method. Alternatively, several companies sell prepared vials containing the requisite chemicals, and the vials can be directly inserted into a colorimeter when the reflux digestion is complete. This is the method we describe here.

Collect samples of natural waters using an appropriate collection device. It would be interesting to compare subsurface samples with samples taken at mid-depth, and also with bottom samples. It would also be instructive to compare samples from the same body of water before and after a rain. In Honolulu, Hawaii, or Seattle, Washington, this would be no problem, but it may be of some difficulty at Death Valley U. However, there are usually local areas with decaying water vegetation that provide good sampling sites.

After collecting samples, the COD test should be run as soon as possible. Otherwise, use the appropriate preservation method described in Experiment 2. The test described here uses Standard Method 5220D (Ref. 2), with specific instructions given in the Bioscience kit.

1. Preheat a COD heater block to 150°C. Care must be taken because the sample vials contain concentrated sulfuric acid. For this reason, the heating block should be used in a fume hood and shielded.

2. Process four standards, one blank, and at least four samples. The blank is prepared by using 1 mL of DI water pipetted into a vial. Several students, or the entire class, can use the same standards. The standard solutions are made by using potassium hydrogen phthalate to give COD equivalents from 20 to 500 mg O_2/L.

3. Prepare the KHP standards using the following procedure.

   a. Dry ground primary standard grade KHP for 2 hr at 120° C.

   b. Prepare a 500 ppm stock solution by dissolving 0.425 g KHP in DI water and diluting to 1.000 L in a volumetric flask. KHP ($HOOCC_6H_4COOK$), has a theoretical COD of 1.176 mg O_2/mg KHP, and the stock solution has a theoretical COD of 500 mg O_2/L.

   c. Prepare 25 mL of each standard having the following equivalents of KHP: 20, 100, and 200 mg/L. These are prepared by pipetting 1, 5, and 10 mL of 500 ppm stock solution into 25 mL volumetric flasks and diluting to volume with DI water. The stock solution is also used as a standard.

4. Use a pipet to transfer 1.00 mL of each sample down the side of a vial containing all the necessary reagents so that it forms a layer on top of the reagents.

5. Seal the cap tightly and mix by shaking.

6. Place the sealed vials in the COD heater block and maintain at 150°C for 2 hours.

7. Remove the vials with a test tube holder, and then transfer to a test tube rack to cool and to allow time for suspended particulates to settle.

8. Wipe the outside of the vials to dry.

9. Set the wavelength on the colorimeter to 600 nm. Use the left-hand knob on the front of the instrument to adjust 0%T with no cuvette in the sample holder. Then use the blank prepared in Step 2 to set the meter or digital readout to 100%T, using the right-hand knob on the front of the instrument.

10. Read the absorbance for each standard and test sample if the instrument is provided with a digital readout. Otherwise, measure the %T for each vial and convert to absorbance [$A = 2 - \log(\%T)$].

Waste Minimization and Disposal

1. The environmental water samples may have been used in previous experiments, and if there is enough left after this experiment, they can be used in several subsequent experiments. Otherwise, they can be disposed of in the laboratory drain if they are not considered to be hazardous waste (if the instructor so directs). If they are hazardous waste, they will be disposed of according to the directions of your instructor.

2. Any extra KHP solid should not be returned to the stock bottle, but placed in a designated container. The standard KHP solutions can be disposed of in the laboratory drain (make sure your instructor approves).

3. The sample vials contain potassium dichromate, silver ion, mercuric ion, and concentrated sulfuric acid. These **cannot** be disposed of in the laboratory drain! The vials will be collected for the entire class and the wastes will be given to a commercial hazardous waste–disposal facility.

Data Analysis

1. Prepare a table of absorbances and concentrations for each standard and sample.

2. Prepare a calibration curve of absorbance (ordinate) versus mg O_2/L (abscissa). Carry out a least-squares analysis on the data, using a computer. Report the correlation coefficient, the slope and the *y*-intercept.

3. Use the calibration curve (or the equation of the previous straight line) to determine the COD of each sample.

4. Discuss your COD values as fully as possible. What substances contribute most to its value? How does it compare with other known values?

5. Using your COD results, estimate the corresponding BOD values.

6. List three possible sources of error in this experiment. How might they be eliminated?

Supplemental Activity

1. If possible, collect water from storm drains before and after a rain and measure their COD. Stormwater runoff is a major contributor to the pollution of streams.

2. Digest a sample of an organic compound, such as ethanol, after appropriate dilution, to see if the experimental COD is equal to its theoretical value.

Questions and Further Thoughts

1. It may be of interest to spike a DI water sample with beer, or other alcoholic products, and determine the COD of the sample. Milk has been suggested as a good sample for study. Other foods and drinks we pour down kitchen drains that eventually enter waterways might be of interest to the student. However, care must be taken to sufficiently dilute the sample to avoid exceeding the capacity of the COD vial to oxidize the sample.

2. Why must the suspended material in the vial be allowed to settle before measuring the absorbance?

3. What chemical species is being measured at 600 nm?

4. Can you balance the equation for the oxidation of KHP? If you can, show the relationship between BOD and COD and why 1 mg of KHP is equivalent to 1.176 mg O_2.

Notes

1. Some companies sell vials that fit directly into colorimeters, such as the Spectronic 20. If the vials do not fit directly into your instrument, simply transfer to a cuvette, being careful to decant to avoid disturbing settled particulates.

2. The vials contain potassium dichromate, concentrated sulfuric acid, mercuric sulfate, and a low concentration of silver ion.

3. Companies that supply kits for the COD analysis include Bioscience, Fisher Scientific, and HACH.

Literature Cited

1. C. Baird, *Environmental Chemistry,* Freeman, New York, 1995.

2. A. E. Greenberg, L. S. Clesceri, and A. D. Eaton, Eds., *Standard Methods for the Examination of Water and Wastewater,* 18th ed., American Public Health Association, Washington, DC, 1992.

Organic Chemical Properties of Natural Waters and Wastewaters

3

Experiment 14

Determination of Oils and Greases by Soxhlet Extraction

Objective—The objective of this experiment is to illustrate the recovery of hydrocarbons from various environmental samples using Soxhlet extraction.

Introduction—Extraction methods are based on differences in solubilities of compounds in the extraction media. Liquid-liquid extraction uses two immiscible liquids brought into close physical contact either through shaking, as in a separatory funnel extraction, or through lengthy reflux of the sample with an appropriate solvent. The analyte, contained in one liquid, distributes between the two solvents until equilibrium is established. Equilibrium depends on the relative solubility of the analyte in the two solvents. In its simplest form, liquid-liquid extraction involves shaking the two solvents in a separatory funnel and allowing the two phases to separate. The phrase "like dissolves like" applies to this situation; the concentration of the analyte is higher in the solvent with polarity characteristics similar to the analyte.

The concentration of an analyte in a liquid phase depends on its solubility in that liquid. The amount of solute extracted depends on the volume of the liquid used. Logically, the more solvent used, the more solute (analyte) will be extracted from the other phase.

Amount of analyte extracted = Concentration of analyte x Volume of solvent

Even this simple example is more complex than indicated. In practice, factors such as temperature, surface area, and time of contact between the two phases may also be important.

Similar techniques can be applied to the extraction of solutes from a solid sample into a liquid phase. Special micropore membranes can also remove certain analytes from liquid phases.

A typical application of extraction is the transfer of a species from one solvent into another. This is known as "solvent extraction." For example, the purification of a drug can be accomplished by either removing the desired species into a new phase or removing interfering substances without extracting the desired material. This is typically done in a "batch mode," with one extraction performed at a time. Mathematically, the distribution of a solute between two phases can be described by a distribution ratio, D, such that,

$$D = [X_{\text{liquid 1}}] / [X_{\text{liquid 2}}] \qquad (14\text{–}1)$$

where liquid 1 is the liquid into which the solute is extracted, liquid 2 is the original liquid, and the brackets indicate concentrations in mol/L.

For most systems, a single extraction is not usually sufficient. A solute that is only sparingly soluble in the extracting liquid (poor choice of solvents) will dissolve to only a small extent before the solvent is saturated. The fraction extracted, θ, can be calculated if D is known:

$$\theta = C_1V_1 / (C_1V_1 + C_2V_2) = DV / (1 + DV) \qquad (14\text{-}2)$$

where C_1 and C_2 are the concentrations in the two phases, V_1 and V_2 are the volumes of the two phases, D is the distribution ratio, and $V = V_1 / V_2$.

For simplicity, consider the case where the volumes of the two phases are equal. In this case, $\theta = D / (D + 1)$. The fraction of original material remaining after one extraction is $(1 - \theta)$. If this solution is then extracted in the same manner, after two extractions the fraction remaining will be $(1 - \theta)^2$ and after n extractions, $(1 - \theta)^n$.

If, for example, only 1% of the solute is extracted each time, more than 450 extractions are required to reduce the concentration by 99%. Thus even in this case there will be some solute remaining in the original phase!

Soxhlet Extraction

As the previous example shows, there can never be 100% extraction of any substance. Some solute will always be left behind, although it may reach negligible amounts at some point. It is up to the analyst to decide when an extraction is "finished." Quite often, the distribution ratio, D, is not known for a specific analyte so that the number of extractions must be approximated. In order to simplify the extraction process, continuous extraction methods have been developed. One widely used technique, and the one recommended by the EPA for extracting hydrocarbons in sediment, fish, or air particulate samples, is Soxhlet extraction.

A typical Soxhlet apparatus is shown in Figure 14-1, although the exact form of the glassware depends on the size of the sample being extracted (which can vary from milligrams to kilograms). In general, the apparatus consists of three glass sections—lower, middle, and upper—which are shown in the figure.

Samples extracted with a Soxhlet usually contain a mixture of solutes. These mixtures must be further separated to quantitate individual species in the mixture. These separations can be accomplished with gas or liquid chromatography. For many purposes, however, a crude single-step extraction can be used to estimate the total amount of analyte in an environmental sample.

Figure 14-1 Soxhlet Extraction Set-Up

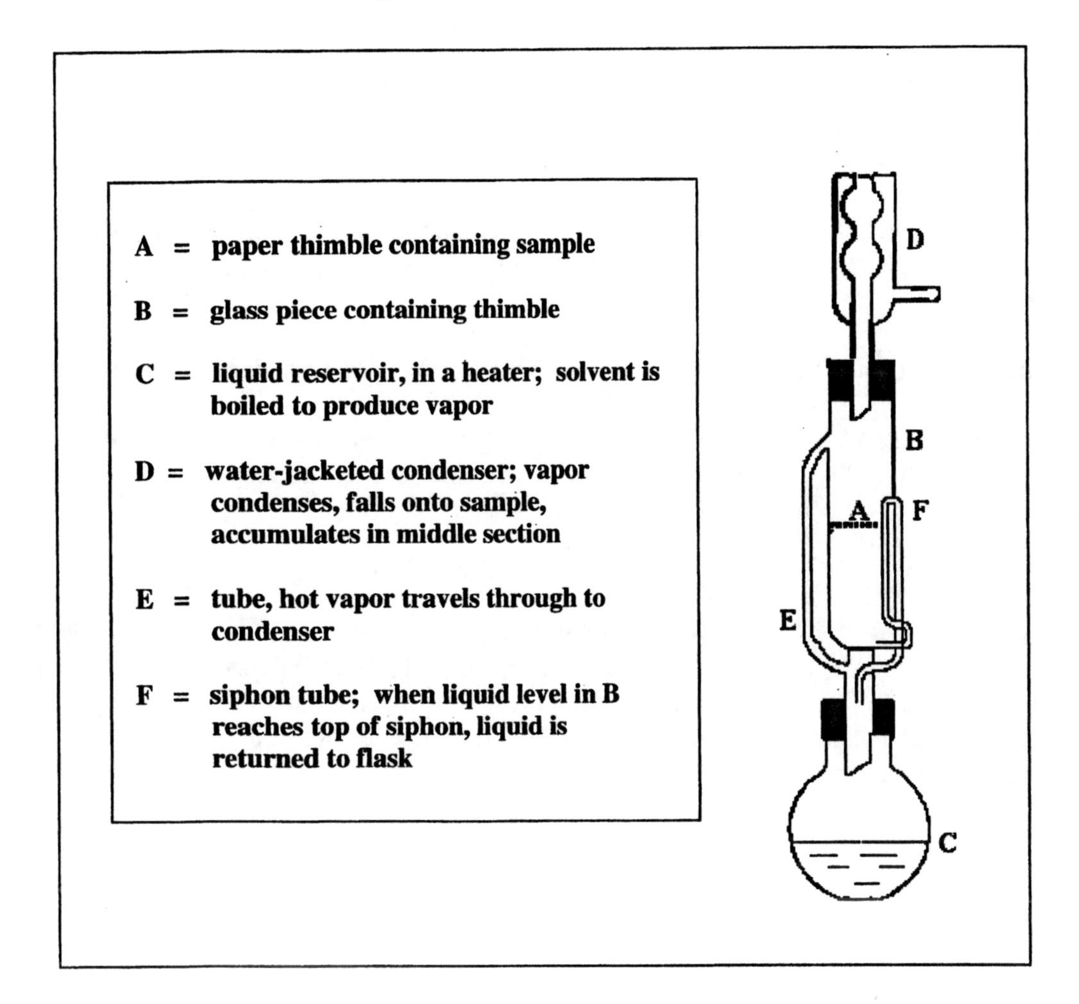

Oils and Greases

Oils and greases are made up primarily of fatty matter of animal and/or vegetable origins and also hydrocarbons of petrogenic (petroleum) origin. Thus, "oils and greases" are a class distinguished on the basis of its solubility characteristics, not its chemical composition. However, they are characterized as having similar physical characteristics, such as solubilities. Oils and greases are nonpolar organic substances that are very soluble in nonpolar organic solvents such as the *n*-alkanes, benzene, toluene, freons, chloroform, carbon tetrachloride, and dichloromethane.

This class of substances is environmentally important because of the large number of sources and their toxicity to the environment. They are present in sewage outfalls, stormwater runoff, industrial discharges, and result from spillages in the off-loading of petroleum. In some states the regulations on oils and greases are "zero discharge." This is due to the fact that very small quantities can spoil vast amounts of clean water.

Safety Issues

1. Safety glasses must be worn at all times in the chemistry laboratory.

2. Great care must be exercised in shaking and venting a separatory funnel. A base neutralizer must be available in case of an accident.

3. When using the Soxhlet, the room must be well ventilated and the water supply to the condenser must be stable. The condenser must be long enough to contain the vapors. For added safety, the extraction can be done in a fume hood.

4. Gloves must be worn when handling hydrocarbon extracts and internal standard solutions (if used).

5. If flame-ionization gas chromatography is used to analyze the hydrocarbon extract, an instructor must give directions on setting-up, using, and shutting down the gas chromatograph before you use the instrument.

Procedure—Although hydrocarbons are ubiquitous in nature, they are present only at very low ppm levels in pristine areas. In urban areas, they can be found in river sediments where there is industrial activity. Also, they can be found at elevated levels in areas near marinas. In areas around gasoline pumps, sediments can be found that provide good study samples. If a good environmental source cannot be found, scrapings from a driveway will suffice.

Samples should be stored in glass at 4°C. Since many bacteria degrade petroleum, samples should be extracted soon after collection. The following procedure applies to the analysis of a stream sediment.

A: Sample Preparation

1. Pre-extract cellulose thimbles (for Soxhlet extraction) in a bottle containing 75 mL hexane and 100 mL of methanol. Soak for at least one hour to extract the hydrocarbons that are at low but detectable levels. Remove the thimbles with forceps and allow excess solvent to drain. Then, holding the thimbles with forceps, rinse with pesticide grade hexane. Any solvent and rinsing wastes must be placed in a designated waste receptacle. The thimble is dried in an oven at 120°C for 0.5 hour.

2. If feasible, the sediment sample should be homogenized in a 1 L Waring-type blender to achieve a uniform sample for analysis. Decant excess water from the homogenized sample and transfer about 45–55 g (weighed to 0.1 mg) to the clean and dry thimble, which has been tared on an analytical balance. The thimble can be supported in a plastic beaker. The thimble should not be filled more than two-thirds full.

B: Preparation of the Internal Standard

1. If gas chromatography is not going to be used for the final analysis, this section can be omitted. Prepare an internal standard solution by weighing out 80 mg of androstane and 40 mg of *ortho*-terphenyl. Quantitatively transfer the standards to a 100 mL volumetric flask, using methanol to transfer and dilute to the mark.

2. After thorough mixing, pipet 10.00 mL of the solution to another 100 mL volumetric flask, fill to the mark with methanol, and thoroughly mix. (This may be provided.)

C: Soxhlet Extraction

1. Place 75 mL of pesticide-grade hexane, 50 mL of pesticide-grade methanol, 10 ml of nanopure water, 10 g of high-grade potassium hydroxide pellets, and several Teflon boiling chips into a 250 mL round bottom flask with a ground glass joint.

2. Place the flask in a heating mantle connected to a Variac and connect the flask to the middle section of the Soxhlet apparatus.

3. Use a forceps to transfer the filled thimble to the middle of the Soxhlet apparatus. Use a pipet to transfer 0.500 mL of the internal standard solution to the top of the sediment in the thimble.

4. Connect the condenser to the middle part of the Soxhlet, turn on the water supply, turn the Variac on, and adjust it to achieve reflux in the condenser. Continue refluxing for a minimum of 12 hours.

D: Sample Clean-Up

1. Turn the heating mantle off and allow the apparatus to cool to room temperature. Because of the volatility of hexane and methanol, it is important that the solution in the flask be at room temperature before proceeding. Transfer the contents of the boiling flask to a 500 mL separatory funnel (with a Teflon stopcock). Rinse the flask with three 15 mL portions of hexane and add the rinsings to the separatory funnel.

2. Stopper the funnel, turn upside-down, and vent. Gently shake, invert, and vent again. Repeat until only a small amount of gas is vented after shaking. The funnel can then be shaken more vigorously. Shake and vent two additional times. Place the funnel in an upright position in a ring on a ring stand. Allow phase separation to occur.

3. Now drain the bottom layer into a 250 mL beaker and retain. Drain the top layer into another 250 mL beaker, and also retain.

4. Return the bottom layer to the funnel, rinse the beaker with two 25 mL portions of hexane, adding the rinsing to the separatory funnel. Stopper the funnel, invert, and vent. Gently shake, invert, and vent again. Shake more vigorously, invert, and vent.

5. Allow the phases to settle, then draw off the bottom layer into the same catch beaker you used previously. Combine the top layer with the hexane extract from Step 3.

6. Repeat Steps 4 and 5 one more time. After this last extraction the bottom layer is drawn off and discarded into an assigned waste container. The top layer is combined with the previous hexane extracts. Keep the combined hexane extracts, which contain the hydrocarbons.

7. Return the combined hexane extracts to the funnel. Add 40 mL saturated aqueous sodium chloride, stopper the funnel, invert, and vent. Shake and vent three times and then allow the phases to separate.

8. Discard the bottom layer using the same container as was used previously. Repeat the clean-up procedure one more time with saturated NaCl, again discarding the bottom layer appropriately.

E: Drying and Evaporation of the Hexane Extract

1. To the hexane extract add enough anhydrous sodium sulfate to lightly cover the bottom of the beaker. Cover the beaker with a watchglass and dry the extract by stirring with a magnetic stir bar and plate for 0.5 hour, working in a hood due to the volatility of the hexane.

2. Filter the extract using a medium porosity filter paper (such as Whatman Number 1) into the smallest beaker that is large enough to contain the sample. Rinse the extract beaker two times with small portions of hexane, collecting the rinsings in the same beaker.

3. Place the beaker in a fume hood with an aluminum foil tent over the top, and allow the solvent to evaporate overnight (this will probably not be to dryness, but if so, add a few mL of hexane). When the sample volume approaches 5 mL, rinse down the sides of the beaker with hexane (which is replacing methanol).

4. Rinse a clean 20 mL scintillation vial three times with hexane. Dry at 110°C in a drying oven and cool to room temperature. If the hydrocarbon analysis is to be gravimetric, weigh the vial. Transfer the sample from the beaker to the vial and place the vial in a small beaker to prevent its being knocked over. Rinse the beaker with three small portions of hexane and add the rinsings to the vial. Continue the evaporation.

Analytical Options

A. Gravimetric Analysis

1. For gravimetric analysis, evaporate the sample to dryness at ambient temperature. (Volatile hydrocarbons will be lost, but most have probably been previously lost due to evaporative and solubility processes.)

2. Reweigh the vial to obtain the total mass of hydrocarbons in the original sample plus the mass of added internal standard.

B. GC/FID Analysis of the Hydrocarbon Mixture

1. Evaporate the sample to a final volume of 2–3 mL.

2. The hydrocarbon mixture can be analyzed by a gas chromatograph equipped with a flame ionization detector, according to instructions given in Experiment 18. This method separates the mixture into individual components.

3. Both the androstane and *ortho*-terphenyl suffice as the internal standards if quantitation is the goal. This is based on a direct relationship between peak area and the mass.

4. Comparing the total hydrocarbon area to the area of the internal standards allows the total hydrocarbon mass to be calculated.

C. Separation and Analysis of Aliphatic and Aromatic Hydrocarbon Fractions

1. This procedure goes a step farther than the original extraction of hydrocarbons. In this extension we separate the hydrocarbon fraction into aliphatic and aromatic hydrocarbon sub-fractions by using gravity column chromatography according to instructions given in Experiment 18.

2. When the two fractions have been prepared, several analyses are possible:

a. Follow the instructions for preparing an extract for gravimetric analysis (preweigh two hexane-washed scintillation vials, evaporate each hydrocarbon fraction to dryness and reweigh the vials). One fraction gives the total aliphatic and the other the total aromatic hydrocarbons.

b. Analyze the aliphatic and aromatic fractions separately by FID/GC as described in Experiment 18. The internal standard is androstane for the aliphatics and ortho-terphenyl for the aromatics.

c. Use the separated aromatics fraction and follow the instructions in Experiment 15 to determine the polycyclic aromatic hydrocarbons (PAHs) fluorimetrically. This analysis can also be done on the unseparated extract.

Waste Minimization and Disposal

1. Several hazardous wastes are generated in this experiment, and efforts should be made to minimize these. The sample(s) collected is(are) contaminated with hydrocarbons, and any extra sample must be properly disposed of. Thus a minimum amount of sample is collected for analysis. Any sample not used will be collected by the instructor and will be disposed of by a commercial hazardous waste facility unless your laboratory has an alternative acceptable procedure.

2. The androstane and *ortho*-terphenyl internal standard solution can be saved and used for future classes. Otherwise, disposal by a commercial facility is recommended.

3. Toluene and hexane wastes will be generated and should be collected in a glass bottle until disposal.

4. Potassium hydroxide should be carefully neutralized with dilute hydrochloric acid. The heat generated can cause volatilization of hydrocarbons and the neutralization should be done in a hood.

5. The final extracted hydrocarbons will be present in very small amount and can be combined with extracted hydrocarbons from subsequent classes. Otherwise, they can be combined with the hexane and toluene wastes that will be disposed of by a commercial facility.

Data Analysis

1. If only a gravimetric measurement of total hydrocarbon content was done, take the measured mass of hydrocarbon (in mg) and divide by the total sample mass (in kg). This gives a result on the basis of wet weight. If a dry weight basis is desired, use the instructions given in Experiments 2, 9, or 10 for obtaining the dry weight.

2. If a gas chromatographic analysis is performed without prior separation into aliphatic and aromatic fractions, calculate the total mass of hydrocarbon by using the area and mass of the internal standard. The mass divided by the area of the internal standard is multiplied by the total area to give the total mass. This mass (in mg) divided by the sample mass (in kg) gives the ppm of hydrocarbon in the original sample. This should be calculated two times, once for each of the internal standards.

3. If the hydrocarbon extract is separated into aliphatic and aromatic fractions, the following may be done:

   a. If the fractions are evaporated to dryness, the masses of the two fractions may be converted to mg/kg as in Part 2.

   b. If the fractions are analyzed separately by gas chromatography, each fraction is quantitated by using the internal standards for each fraction, androstane for the aliphatics and *ortho*-terphenyl for the aromatics. Report each fraction in mg/kg of sample (either wet or dry weight of sample). Compare the aliphatics mass to the aromatics mass. This ratio is characteristic of each hydrocarbon type.

   c. Use the fluorescence intensity of the aromatics fraction to calculate the equivalent mass of anthracene per kg of sample.

Supplemental Activity

1. If two or more groups are working simultaneously, the use of different sample types can be very instructive. Comparing a sample from a pristine area with one from an industrial area (a shipyard is good, or a military base) often produces large differences in results.

2. It would be interesting to examine the aromatics fraction by using ultraviolet spectroscopy.

Questions and Further Thoughts

1. By comparing the gas chromatograms obtained with those for standards, provided by the instructor (or determined by the student), it is possible to identify specific compounds, such as the *n*-alkanes or simple aromatics.

2. Why are plasticizers, such as dibutylphthalate or dioctyladipate, extracted with hydrocarbons; further, if the hydrocarbon fraction is separated into aliphatic and aromatics fractions, why do these plasticizers preferentially end up in the aromatics fraction?

3. Why are the phthlate and adipate plasticizers so common in the environment?

Notes

1. Oil-spill samples are particularly interesting to study. Areas where oil spills have recently occurred can be indicated by the U.S. Army Corps of Engineers and the U.S. Coast Guard.

2. It is noteworthy that the plasticizer content of samples may be so large that it interferes with the separation and quantitation of individual components.

3. When working with hydrocarbon samples, particularly when very sensitive analytical methods are used, it is critical that extreme care be taken to avoid contaminating the samples or the extracts. This is also true for most environmental samples and procedures since methods are generally designed to provide very low detection limits. Minute contaminants in solvents, reagents, absorption media (silica gel, alumina, etc.), and on the surfaces of sample containers and processing equipment can often render results unusable. The student must learn to work meticulously to avoid these pitfalls.

Literature Cited

1. A. E. Greenberg, L. S. Clesceri, and A. D. Eaton, *Standard Methods for the Examination of Water and Wastewater,* 18th ed., American Public Health Association, Washington, DC, 1992.

2. Final Report, *Survey of Hydrocarbons in the Lower St. Johns River in Jacksonville,Florida*, D. N. Boehnke, C. A. Boehnke, K. I. Miller, J. S. Robertson, and A. Q. White (prepared for The Florida Department of Natural Resources, Tallahassee, FL, October 1983).

Experiment 15

Fluorimetric Determination of Polycyclic Aromatic Hydrocarbons

Objective—The objective of this experiment is to learn the use of an extremely sensitive method for quantitatively determining polycyclic aromatic hydrocarbons (PAHs). Also, the relative response of three aromatic hydrocarbons will be measured.

Introduction—Earlier in these experiments we used analytical methods based on the absorption of electromagnetic radiation, namely colorimetry and atomic absorption. We now use a method based on the emission of radiation, fluorimetry. This method can be extremely sensitive for certain types of compounds, detecting less than ppm levels, but very insensitive for other types. Also, it is capable of detecting one of many substances present in a mixture without interference.

Theory—The term fluorescence refers to the emission of radiant energy by a molecule or ion in an excited state. The particle reaches the excited state by absorbing radiant energy, usually from the ultraviolet region of the spectrum. The process may be represented by the scheme shown in Figure 15-1.

Figure 15-1 Diagram of Stepwise Process of Fluorescence

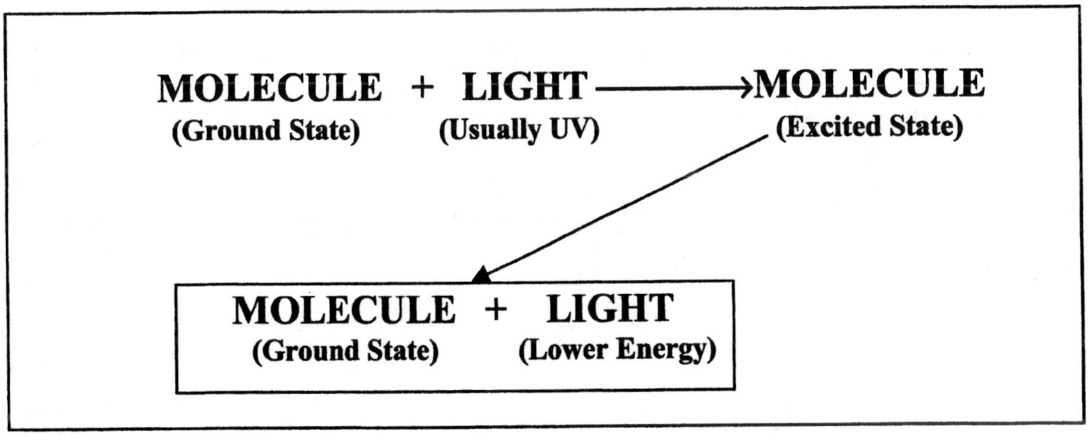

MOLECULE (Ground State) + LIGHT (Usually UV) ⟶ MOLECULE (Excited State)

⟶ MOLECULE (Ground State) + LIGHT (Lower Energy)

After excitation, the species quickly returns to the lowest vibrational sublevel of the first excited electronic state. The energy lost in this process is transferred to solvent molecules and may ultimately appear as heat.

Fluorescence occurs when a species from the ground vibrational level of the excited electronic state emits electromagnetic radiation. If the species becomes deactivated and returns to the ground state by another method, fluorescence is not observed. Deactivation can occur through collisions with molecules of solvent or other fluorescing species, resulting in decreased vibrational energy and finally a transition to the ground electronic state. The more flexible a molecule or ion, the easier it is for collisional deactivation to occur. Thus, in most cases only rigid species will fluoresce.

Most substances that fluoresce strongly enough to be useful analytically are organic rather than inorganic. In general, substances that absorb strongly in the UV, usually at wavelengths longer than 250 nm, may fluoresce. Substances that absorb at shorter wavelengths are often subject to photodecomposition rather than fluorescence because of the high energy of the radiation absorbed.

Many aromatic compounds absorb strongly in the 250–320 nm region. Generally, the substances most likely to fluoresce are those with cyclic structures containing conjugated double bonds. Most polynuclear aromatic hydrocarbons fluoresce strongly. The presence of –OH, –OR, or $-NH_2$ groups increase fluorescence intensity. (R is an organic substituent, such as methyl, ethyl, phenyl, etc.) Groups such as –COOH, $-NO_2$, or $-SO_3H$ reduce or even eliminate fluorescence. Ring closure in a compound, as in the formation of a metal chelate, enhances fluorescence.

Different chemical species have characteristic excitation and fluorescence spectra. In order to measure fluorescence, it is necessary to have a source that emits UV radiation, a detector that responds to UV and visible radiation, and a means for separating the excitation and fluorescence radiation.

The experimental arrangement for fluorimetry is shown in Figure 15-2. The excitation radiation is produced by a high-pressure xenon lamp that produces light over a wide range of wavelengths. This radiation then passes through a narrow-bandpass primary filter to give the proper energy range for exciting the species under study. As the sample fluoresces (in all directions) the emitted radiation is measured at right angles to the excitation beam by the detector after it passes through a sharp-cut secondary filter. The purpose of the secondary filter is to remove unwanted (for example, scattered) incident radiation.

As mentioned, most fused-ring aromatics fluoresce after UV excitation. The absorbance spectra of fluorescent species often closely resemble the excitation spectra, both of which are generally broad band. Thus, a single excitation wavelength may excite several potential analytes. Fortunately, the emission spectra of excited species differ from the excitation in ways that are determined by the structure of the molecule involved. Often a proper selection of excitation and emission wavelengths will permit a selective analysis for a particular substance. The correct selection of excitation and emission wavelengths is often determined using a spectrophotofluorimeter.

Although basically dissimilar to colorimetry, fluorimetry obeys a "Beer's-type" law,

$$F = k'C \tag{15-1}$$

where F is the intensity of emitted fluorescence, k' is a constant that depends on several factors, and C is the concentration. As with colorimetry, a calibration plot is prepared using a series of standard solutions and measuring their fluorescence intensity.

Figure15-2 Diagram of a Filter Fluorescence Photometer

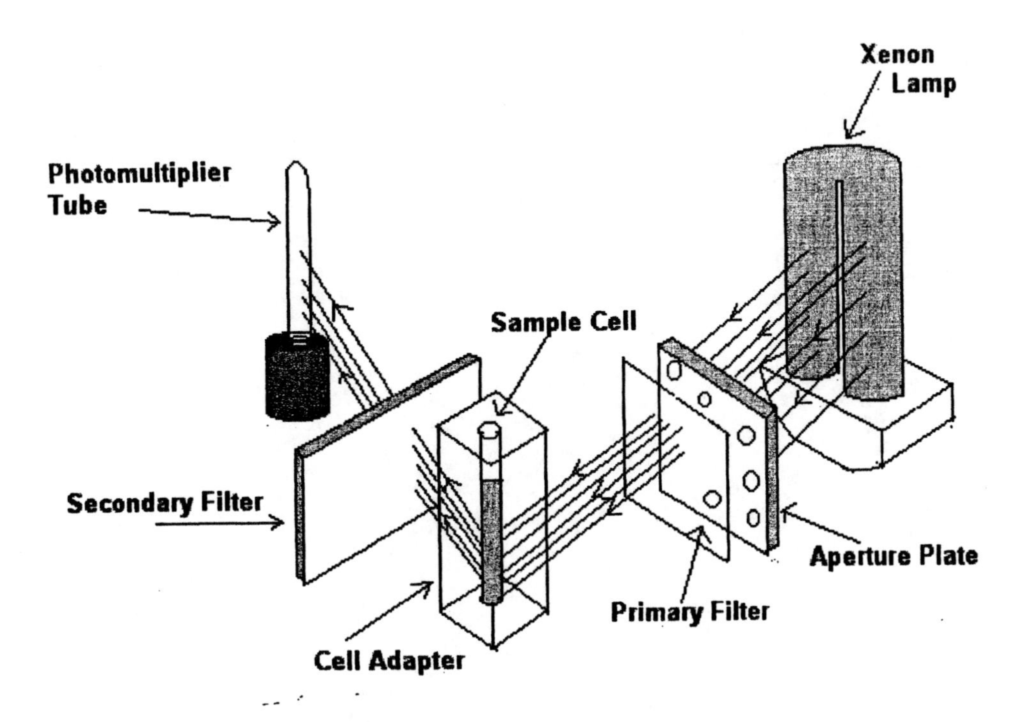

Polycyclic Aromatic Hydrocarbons One of the most useful applications of fluorescence is the determination of aromatic hydrocarbons. Since aromatic hydrocarbons are among the chief air pollutants, the fluorimeter has been used by the Public Health Service for determining these substances. The measurement of the very low concentrations of PAHs found in samples of particulates collected from the atmosphere is among the usual applications.

Aromatic compounds are characterized by a ring structure having alternating single and double bonds, and benzene is the simplest example. Benzene itself fluoresces only slightly due to its vibrational flexibility. When a second ring is attached to the first to give napthalene, which is quite rigid, there is intense fluorescence. A third ring can be attached to the previous ring system in two ways, one giving a linear arrangement called anthracene, and a third nonlinear arrangement called phenanthrene. These substances are illustrated in Figure 15-3. Compounds with fused rings are called polynuclear aromatic hydrocarbons (PNAs) or polycyclic aromatic hydrocarbons. They have a planar geometry that gives them a rigidity, which results in intense fluorescence.

PAHs originate from both biogenic (natural) and anthropogenic (human) sources. The incomplete combustion of hydrocarbons is the chief means of PAH formation. Among the PAHs found in urban air are naphthalene, acenaphthene, acenaphthylene, fluorene, phenanthrene, anthracene, fluoranthene, pyrene, benzo(*a*)pyrene, benzo(*a*)anthracene, chrysene, and many others.

PAHs are present at low levels in automobile exhaust. Higher PAH levels occur when large amounts of soot are also present, as in diesel exhaust or smoke from coal or wood fires. Atmospheric PAHs are found almost exclusively in the solid phase, usually sorbed to soot particles. PAHs are produced from saturated hydrocarbons at high temperatures under oxygen-deficient conditions. Elevated PAH levels are likely to be encountered in polluted urban air and in the vicinity of forest fires. Coal furnace stack gas may contain 1000 $\mu g/m^3$ PAHs.

PAHs are formed on the surface of charred or burnt foods. Charcoal broiled and smoked meat and fish contain some of the highest PAH levels found in food. However, leafy vegetables, such as lettuce and spinach, can be even greater sources of PAHs due to their deposition from the atmosphere onto the leaves of these vegetables.

Figure 15-3 Polycyclic Aromatic Hydrocarbons

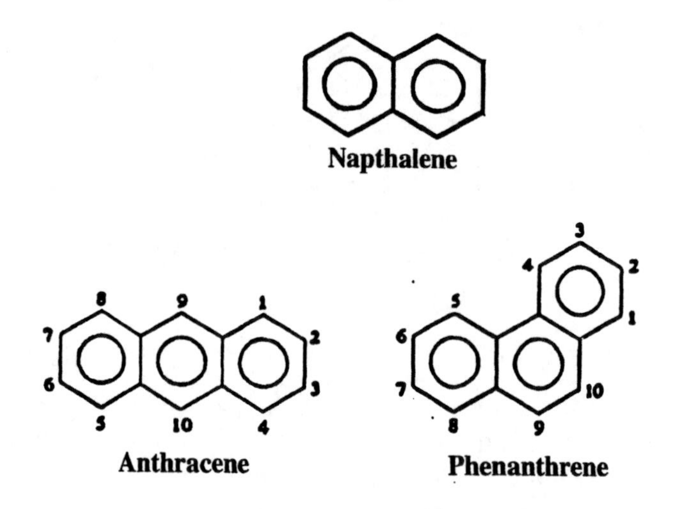

The chemical composition of tobacco smoke is complex and contains thousands of components, many of which are carcinogens. Environmental tobacco smoke consists of gases and particulates. The gases include CO, NO_2, HCHO, PAHs, and other volatile organic compounds (VOCs). The particulate phase is called tar, which contains nicotine and the less volatile hydrocarbons.

Cigarette smoke contains about 100 $\mu g/m^3$ PAHs. The concentration of toxic products of partial combustion is actually higher in second-hand smoke than in mainstream smoke since combustion occurs at a lower temperature in a smoldering cigarette than in one through which air is inhaled.

The presence of naturally occurring PAHs is an interesting feature of soil organic matter. PAHs found in soil include fluoranthene, pyrene, and chrysene. In rivers and lakes, anthracene and phenanthrene are found attached to sediments rather than dissolved in the water. PAHs containing four or fewer rings usually exist in the gaseous state when released into the atmosphere. After less than one day they are degraded by free radical reactions since PAHs are light-sensitive.

Compounds with four or more benzene rings fused together can be potent carcinogens. Their cancer-causing capacity is due to activation by the same class of liver enzymes that metabolizes toluene and other xenobiotics. When these enzymes add oxygen to the PAHs, they produce epoxide adducts that interact strongly with the bases of DNA to alter genes. The PAHs that are the most potent carcinogens possess a bay strongly with the bases of DNA to alter genes. The PAHs that are the most potent carcinogens possess a bay region formed by a branching of the benzene ring sequence.

PAHs enter the aquatic environment as a result of oil spills from tankers, refineries, and offshore drilling. Also, the leaching of PAHs from the creosote used to preserve the immersed lumber of fishing docks represents a significant source of pollution to crustaceans such as lobsters. PAHs have been found to bioaccumulate in the fatty tissues of some marine organisms, and have been linked to the occurrence of liver lesions and tumors in some fish.

High levels of benzo(*a*)pyrene have been found in the sediments at several urban sites in the Great Lakes. It bioaccumulates in the food chain (log K_{ow} = 6.3; see Experiment 16). Methyl-substituted PAHs are often more potent carcinogens than the parent hydrocarbons.

The PAH levels in drinking water are usually less than a few ng/L, an insignificant level for humans.

Safety Issues

1. Safety glasses must be worn at all times in the chemistry laboratory.

2. Wear gloves when handling aromatic hydrocarbons. Be careful to avoid skin contact with these substances.

3. Do not pour any unused reagents down the drain, but dispose of wastes according to your instructor's directions.

Procedure

1. **Preparation of Calibration Curve:** A 1.0 x 10^{-3} M stock solution of anthracene (17.8 mg/100 mL cyclohexane) will be provided. Transfer 1.0 mL of the stock solution to 10, 25, 50, and 100 mL volumetric flasks and dilute to the mark with cyclohexane. Use cyclohexane as the blank, and determine the fluorescence intensity of each standard using the operating procedure for your instrument.

2. Prepare 100 mL of 1 x 10^{-2} M solutions of toluene, naphthalene, and biphenyl using cyclohexane as solvent. Measure the fluorescence intensity of each solution.

3. Extract a used cigarette filter by placing it in a beaker with about 25 mL of ethanol (the ethanol provides both hydrophilic and hydrophobic interactions to extract the hydrocarbons in an aqueous environment). Stir for 0.5 hour. Dilute 1 mL of the extract with enough ethanol to give 25 mL of solution. Measure the fluorescence intensity of this solution. If the fluorescence intensity is not in range, adjust the dilution appropriately.

Waste Minimization and Disposal

1. A considerable volume of hazardous waste can be generated in this experiment. Efforts should be made to minimize the wastes. The calibrating solutions of anthracene in cyclohexane can be shared to minimize the wastes due to this source.

2. The toluene, naphthalene, and biphenyl solutions can be combined and put into a bottle for later commercial disposal. The solution of anthracene in cyclohexane can also be combined with the other aromatic wastes.

3. The alcohol extract of a cigarette filter should be put into a designated container for later disposal by a commercial hazardous waste facility.

Data Analysis

1. Prepare a calibration curve of fluorescence signal (as ordinate) versus concentration of anthracene. Compare with the plot shown in Figure 15-4. Estimate the lowest concentration of anthracene that could be accurately measured by fluorimetry. How does it compare with UV spectroscopy? The curvature in Figure 15-4 is due to self-absorbance by the anthracene when it is present at high concentrations.

2. For the test samples, prepare a table of fluorescence intensity and concentration, as determined from the calibration plot.

3. Compare the fluorescence intensity of toluene, naphthalene, and biphenyl. Discuss the effect of structure on fluorescence intensity. Is it valid to use anthracene to estimate other PAHs?

4. Calculate the "effective anthracene concentration" for the cigarette filter extract. Estimate the mg of anthracene in the filter, assuming that the fluorescence is due entirely to anthracene.

Supplemental Activity

1. If the previous experiment has been completed and a sample of environmental hydrocarbons has been extracted, this may be diluted in cyclohexane and its fluorescent intensity measured to give an estimate of PAHs present in the sample.

2. In a previous experiment the supplemental activity suggests separating aliphatic and aromatic hydrocarbons. This can be carried out for many sample types, including crude oils, or a Soxhlet-extracted sample of oils and greases. This sample can also be quantitated for PAHs.

3. Another important application of fluorimetry is the determination of the concentration of spent sulfite liquor in water. References to this and other applications are given in the Literature section.

4. One of the later experiments in this manual includes the collection of particulates from the atmosphere. These can be extracted for PAHs and then analyzed by fluorimetry.

Figure 15-4 Anthracene Calibration Curve

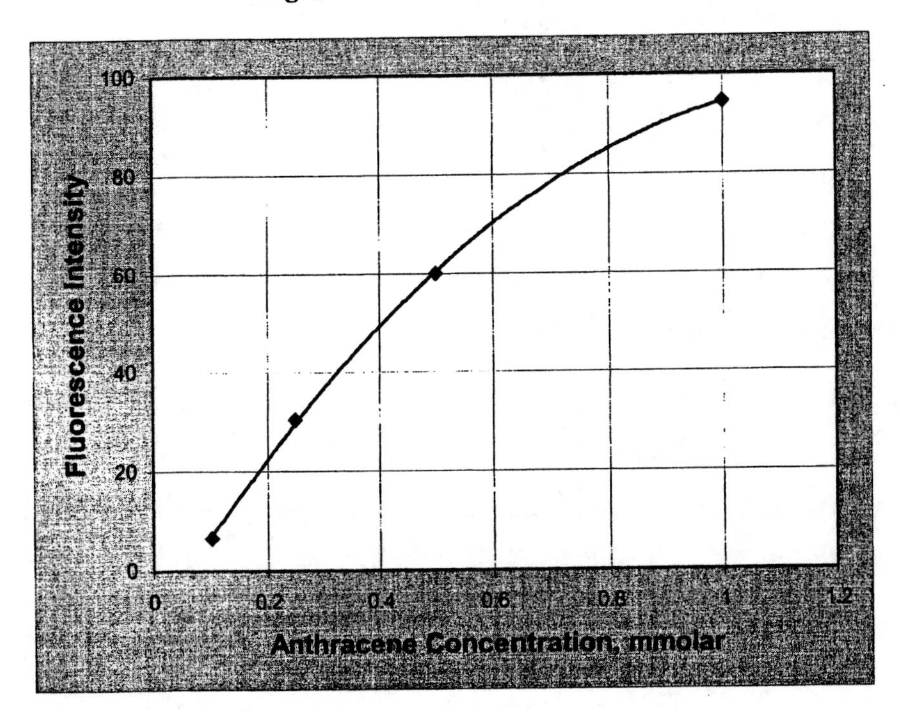

Questions and Further Thoughts

1. Devise a study to show how the difference in the PAH content of second-hand smoke from a cigarette and inhaled smoke could be detected.

2. What are PAHs exposed to in the atmosphere that causes their destruction?

3. Lichens frequently contain fluorescing compounds, and these can be extracted and the total fluorescence intensity determined. Lichens, themselves, are very sensitive to certain air pollutants.

Notes

1. The fluorocolorimeter is a direct-reading filter fluorimeter used for the quantitative determination of substances with known excitation and emission spectra. Prior to using the instrument, the proper filters must be installed. In this experiment, a good choice of primary filter is a PC-6 narrow-pass filter. A PC-9A sharp-cut filter is a good choice for the secondary filter.

2. Before use, the instrument is turned on and allowed to stabilize, the meter is zeroed, and the photometer dark current subtracted from the meter reading. Dark current is a measurable signal present even when there is no light striking the PM tube.

Literature Cited

1. C. Baird, Environmental Chemistry, W. H. Freeman, New York, 1995.

2. R. F. Christman and R. A. Minear, Fluorometric detection of lignin sulfonates, *Trend. Eng Univ. Wash.,19(1)*:3–7 (1967).

3. V. F. Felicetta and J. L. McCarthy, Spent sulfite liquor. X. The Pearl-Benson, or nitroso method for the estimation of spent sulfite-liquor concentration in waters, *Tappi, 46(6)*:337–347 (1963).

4. J. S. Fritz and G. H. Schenk, *Quantitative Analytical Chemistry,* 5th ed., Allyn and Bacon, Boston, 1987.

5. A. E. Greenberg, L. S. Clesceri, and A. D. Eaton, Eds., *Standard Methods for the Examination of Water and Wastewater*, 18th ed., American Public Health Association, Washington, DC, 1992.

6. Stanley E. Manahan, *Environmental Chemistry,* 5th ed., Lewis Publishers, Chelsea, MI, 1991.

Experiment 16

Determination of the Octanol /Water Partition Coefficients for Organic Pollutants of Varying Hydrophobic/Hydrophilic Character

Objective—The objective of this experiment is to show how an important parameter, related to the bioaccumulation of pollutants, the octanol/water partition coefficient, can be measured.

Introduction—An environmental scientist is interested not only in the origin of pollutants, but also in their fate in the environment. The fate of an organic compound depends to a great degree on its volatility. If it finds its way into the atmosphere, where it is subjected to a large concentration of oxygen and sunlight, it can be oxidized or undergo photodecomposition. If it finds its way into the water column of a stream or lake, it will be subjected to dissolved oxygen, a large concentration of water (where it may be hydrolyzed), or bacterial decomposition.

Also, it is of importance to follow the transport of the pollutant. In the atmosphere, this is partly a result of wind and rain. In a stream, transport depends on several factors, including volatility, solubility, and the availability of silt onto which the pollutant can sorb. For example, suppose that 2,4-dichlorophenoxyacetic acid (2,4-D), used as a herbicide, finds its way into a stream. Hydrogen bonding between 2,4-D and water will result in appreciable water solubility. However, the phenyl group gives a hydrophobic interaction with water, limiting its solubility. This factor increases its tendency to sorb to organic matter in the water. For this particular molecule, both hydrophobic and hydrophilic interactions are significant.

If instead of 2,4-D we consider *p*-dichlorobenzene, a common component of mothballs, there is no hydrogen bonding interaction with water, and the molecule will become highly sorbed onto organic matter, if it is present. In this case, the molecule will be transported downstream until the sedimentary material eventually grows to a size where it will settle out. Hydrocarbons from sewage outfalls have been found to be transported only a short distance downstream from their origin before settling out.

For fish, the fate of pollutants is very important. Filter feeders, such as menhaden and flounder, feed from the sediment, and the presence of pollutants sorbed into the sedimentary material can have an adverse effect on these species. On the other hand, water-soluble pollutants can be readily absorbed through gill fins and have an even more detrimental interaction with the organism.

To determine the relative availability of an organic pollutant, an important property is partitioning between the organic (lipid) and aqueous phases. This is quantitated through the octanol/water partition coefficient, K_{OW}.

Theory—The octanol-water partition coefficient is defined as the ratio of the concentration of a compound, X, in the octanol phase, $[X]_o$ divided by the concentration of X in the aqueous phase, $[X]_w$ when the two phases are in equilibrium. Thus,

$$K_{OW} = [X]_O / [X]_W \quad (16\text{-}1)$$

The K_{OW} is used to approximate the bioconcentration factor, BCF. The BCF indicates the degree to which an organism accumulates a pollutant over its concentration in the environment.

Octanol is used to emulate lipid, the fatty substances of an animal organism that, in reality, is a mixture of trialkylglycerates, esters of long-chain carboxylic acids, and glycerol. In this model we assume that fish, in particular, take up pollutants from the water column. Although this may be true for some species of fish, others sorb materials adhering to solid, suspended matter, or from sedimentary materials.

For nonpolar substances such as biphenyl and benz[*a*]anthracene, K_{OW} is 7.6×10^3 and 4.1×10^5, respectively. To facilitate the handling of such large numbers, the logarithm to the base 10 of K_{OW} is used. This is similar to using pH values instead of hydrogen-ion concentrations. For biphenyl, log K_{OW} is 3.88 and for benz[*a*]anthracene it is 5.61. When log K_{OW} is very large, fish cannot extricate the compound from sedimentary material, but if the same substance is in the water column, fish will bioconcentrate it to a great extent.

Smaller, nonpolar molecules, such as benzene and toluene, have much smaller K_{OW} values, log K_{OW} being 2.13 for benzene and 2.69 for toluene. Even when adsorbed onto organic silt, fish can extract such pollutants and bioaccumulate them.

Molecules that have specific interactions with water, especially through hydrogen bonding, have small K_{OW} values and are partitioned more or less equally between both water and octanol phases. Ethanol, for example, has a log K_{OW} equal to -0.31 and chloroform has log K_{OW} equal to 1.95. Such substances tend to undergo little accumulation in organisms. Some octanol/water partition coefficients are given in Table 16-1.

Safety Issues

1. Safety glasses must be worn at all times in the chemistry laboratory.

2. Although pressure may not build up in the separatory funnel, be very careful when shaking and venting the funnel.

Table 16-1 Octanol/Water Partition Coefficients for Various Substances at 25°C

Compound	$\log(K_{ow})$	Compound	$\mathrm{Log}(K_{ow})$
Ethanol	−0.31	Toluene	2.69
Vinyl Chloride	0.60	Chlorobenzene	2.92
Aniline	0.90	*o*-Xylene	3.12
Dichloromethane	1.15	Ethylbenzene	3.15
Benzaldehyde	1.48	Naphthalene	3.59
Nitrobenzene	1.85	*n*-Propylbenzene	3.68
Chloroform	1.95	Anthracene	4.34
Benzene	2.13	Pentachlorophenol	5.01
Carbon Tetrachloride	2.64	*p,p'*-DDT	6.36

Source: H. F. Hemond and E. J. Fechner, *Chemical Fate and Transport in the Environment,* Academic Press, New York, 1994.

Procedure

Summary

To obtain equilibrium K_{OW} values requires mechanically shaking the two-phase system for many hours. This is particularly true when the partition coefficient is very large. Another experimental difficulty can be the very small water solubility of very hydrophobic substances. In this situation, the analysis of the aqueous phase may require considerable sensitivity for large values of K_{OW}. It may be possible to obtain the quantity of hydrophobe in the aqueous phase by taking the difference between the original mass introduced into the system and the amount present in the octanol phase from analysis. That is, we analyze one phase for the hydrophobe and calculate the mass of the substance in the second phase.

To minimize some of the experimental difficulties, we shall examine systems that have relatively small K_{OW} to expedite equilibrium. Also, since the concentrations of the substance will be significant in the aqueous phase, it is possible to analyze just the aqueous phase and avoid dealing with the very messy octanol phase. It will also be possible to use less sensitive methods to determine the concentration of the substance.

Frequently, gas chromatography (GC) is used for the analysis. For chlorinated hydrocarbons, electron capture GC should be used, whereas for hydrocarbons, flame-ionization detection GC is preferable. For sufficiently high concentrations, a thermal conductivity detector can be used. Other methods that can be used for the analysis are UV spectroscopy or fluorimetry (if the substance under study fluoresces). If the concentration is high enough, refractometry is an accurate and convenient method to use.

There are many options for studies of this type. Each group will study three alcohols; ethanol, propanol, and butanol is one such study. Isomers can be examined to observe the effect of branching on K_{OW}.

Experimental

1. Pipet 25.00 mL of 1-octanol into a 150 mL separatory funnel. Record the density of the octanol, which will be on the bottle. Pipet 1.000 mL of sample (that is, ethanol, propanol, or butanol, if this is your system) into the octanol. The density of the alcohol (which may be on the reagent bottle) is used to determine the mass of alcohol added.

2. Pipet 25.00 mL of nanopure water into the funnel, briefly swirl, stopper, and carefully invert and vent through the stopcock.

3. Shake the mixture briefly and vent again. Continue to shake and vent until there is no more pressure release on venting. Shake intermittently for a total of 2 minutes. Put the funnel into a small ring on a ringstand and allow the phases to separate.

4. Repeat the above procedures for two other alcohols.

5. If there is difficulty in phase separation, allow the funnels to sit until the next laboratory period. Alternatively, use a centrifuge to effect separation if a large volume centrifuge is available.

6. Draw off the bottom, aqueous phase, into a small glass-stoppered bottle.

7. Analyze the aqueous phase, using either (a) gas chromatography or (b) refractometry. If GC is used, spike a 5 mL portion of the phase with a known amount of another alcohol, such as methanol or isopropanol. Select the size of the spike so that the spike peak will give an area on the chromatogram approximately equal to the amount expected for your sample alcohol in that phase. For example, if the octanol/water partition coefficient is 3 and if 1 g of alcohol was added to the system, then the amount of alcohol in the aqueous phase would be 0.25 g. The internal standard would be added to that extent (10 mg/mL), using a micropipet or by adding the appropriate amount to the aqueous phase using an analytical balance.

8. If refractometry is used for analysis, make up five solutions of each alcohol in water. For simplicity, make up 25 mL of each solution, spanning the concentrations expected or estimated for your system. For example, if the system has a K_{ow} of 3, as in Part 7, suggested amounts of the alcohol in 25 mL of water would be 0.1, 0.15, 0.25, 0.30, and 0.35 g (accurately measured, to 0.1 mg).

Waste Minimization and Disposal—The only wastes generated in this experiment are alcohols. These can be combined in a waste bottle for later commercial disposal.

Data Analysis

1. If your analysis of the aqueous phase was done using GC, include your gas chromatograms and show how the concentrations of the alcohol in the octanol phase were calculated. In GC, areas are proportional to masses and GC peaks are often triangular, which facilitates area estimation.

2. If your analysis of the aqueous phase was done using refractometry, tabulate standard concentrations (g alcohol/25 mL water) and the refractive indices of the standards. Use a computer to plot the refractive index (as ordinate) versus concentration for the standard solutions. If nearly linear, carry out a least-squares analysis and use either the plot or the equation of the straight line to obtain the concentrations of alcohol in the water phase. If there is curvature, use the plot directly to estimate the concentration.

3. Use your data to calculate log K_{OW} for each alcohol studied. Compare with literature values (see Table 16-1 and references in the Literature Cited section).

4. Discuss the significance of your results. To facilitate discussion, plot K_{OW} versus the carbon number (or the boiling point for the alcohol). Additional data can be obtained from Table 16-1.

a. Why does K_{OW} change from one alcohol to the next?

b. Does the curve extrapolate linearly to octanol (carbon number 8)?

5. Briefly discuss how UV spectroscopy could be used in this experiment if the substances under study were benzene, toluene, ethylbenzene, or the three xylenes? These are important volatile compounds in gasoline that have been found in environmental waters.

Supplemental Activity—An analysis of the aqueous phase by gas chromatography would serve as a check of the results obtained by refractometry.

Questions and Further Thoughts

1. How is bioaccumulation (or the BCF) related to biomagnification?

2. Would the presence of a second alcohol affect the K_{ow} measured for another alcohol? How would it affect the method chosen for the analysis?

Notes—Octanol is very messy to work with and supplemental work should deal only with the aqueous phase.

Literature Cited

1. S. Banerjee, R. H. Sugatt and D. P. O'Grady, A simple method for determing bioconcentration parameters of hydrophobic compounds, *Environ. Sci. Technol.,18(2)*:79–81, (1984).

2. C. T. Chiou, Partition coefficients of organic compounds in lipid-water systems and correlations with fish bioconcentration factors, *Environ. Sci. Technol.,19(1)*: 57–62, (1985).

3. H. F. Hemond and E. J. Fechner, *Chemical Fate and Transport in the Environment,* Academic Press, New York, 1994.

4. D. Mackay, Correlation of bioconcentration factors, *Environ. Sci. Technol.,16*(5):274–277,(1982).

Experiment 17

Determination of Oil-Spill Sources Using Pattern Recognition of Known Petroleum Product Gas Chromatograms

Objectives—There are several objectives to be achieved in this experiment: These include:

1. Introducing you to the basic principles of gas chromatography with thermal conductivity detection

2. Investigating the concept of headspace analysis

3. Observing the pattern of peaks obtained in the gas chromatograms of petroleum and its refined products, and using the patterns to identify the source of environmental hydrocarbons

Introduction—Most common fuels (e.g., gasoline, kerosene, diesel fuel) are mixtures of low-boiling fractions of petroleum. They consist of low molecular weight aliphatic and aromatic hydrocarbons. Solvents, on the other hand, are usually mixtures that may contain aromatics, chlorinated hydrocarbons, aldehydes, and ketones. It is often easy to distinguish between fuels and solvents, but it can be difficult to distinguish between various fuels and between various solvents.

Studies of potential arson cases, hazardous waste incidents, or oil spills require certainty in the results of an analysis, since criminal and civil litigation may result.

The correct interpretation of analyses and the ability to make sound judgments based on qualitative data takes years of experience. Nevertheless, the bases for such judgments are fairly simple. The method of pattern recognition will be used to identify the contaminant, and this experiment will be used to acquaint the student with the requirements of such subjective interpretations.

In this experiment, you are the "expert witness" in the chemical investigation of a spill of an unknown liquid on a beach. A sample of sand containing an organic residue of suspicious origin is taken at the scene. You are asked to analyze the residue. Gas chromatography is used since it is well suited for the analysis of mixtures of volatile organic species. Standard mixtures of several organic chemicals found in fuels and solvents are analyzed to provide chromatograms for comparison. The residue is analyzed by "headspace analysis" to determine the type of organic material present. For safety, the sample is obtained from the residue at room temperature. If the sample is old, or contains high-boiling components, the sample container can be heated by immersing it in hot water to provide more vapor in the headspace; however, the samples used will be sufficiently concentrated in the gaseous phase so they will not require heating. Comparing the chromatograms for known standards with that for the unknown should permit the identification of the fuel or solvent contained in the sample.

A set protocol must be strictly followed so that the retention indices and response factors obtained in the analysis of the standards can be reproduced. Following injection of standards of known composition, samples of the volatile fraction of the residue will be injected into the gas chromatograph.

Theory—Gas chromatography is an instrumental method used to separate a mixture of volatile compounds into its various components. Although it is excellent for separating complex mixtures, it is not good at identifying specific compounds.

The basic components of a gas chromatograph are shown in Figure 17-1. The injection port is used to introduce the sample into the chromatographic system. The column is where the separation of the mixture actually occurs. After eluting from the column, the sample components are carried (via a gas called the "carrier gas") into the detector where they are sensed. Finally, the readout responds to the electrical signal produced by the detector and produces peaks, the areas of which are proportional to the amount of that component in the mixture.

Figure 17-1 Components of a Gas Chromatograph

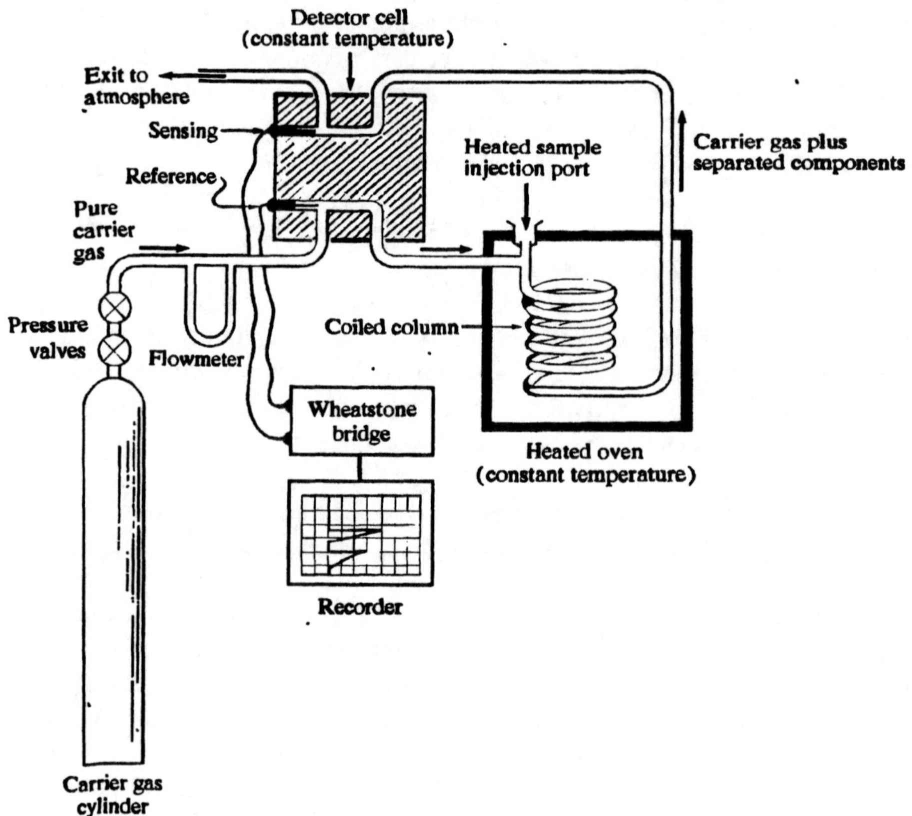

Source: G. H. Schenk, R. B. Hahn and A. V. Hartkopf, *Introduction to Analytical Chemistry*, 2nd ed., Allyn and Bacon, Boston, 1981, p. 424. By permission.

The Injection Port This is where the sample (usually liquid) is introduced into the chromatographic system. The injection is usually done with a microliter syringe that penetrates a rubber-type septum, which provides an airtight system. The temperature of the injection port can be adjusted and is usually held at a temperature such that all the components of the mixture are flash-vaporized when injected.

The Column After the sample is injected, it is carried into the column where the separation of the mixture occurs. There are a number of column types, but we shall examine just one type, a packed column. A packed column is filled with a porous, granular support that is covered with a uniform coating of a nonvolatile liquid, **stationary phase.** The solid support is chosen to have a large surface area, which is needed to support an ample quantity of the coating.

The basis of gas chromatography is the partitioning of components of the mixture between two phases, usually liquid and gaseous phases. The theory for gas-liquid chromatography is very similar to the theory for solvent-solvent extraction as discussed in Experiment 14.

The nature of the stationary phase is chosen to have the same electrical character as the components of the mixture to be separated. If a mixture of hydrocarbons is to be separated, the stationary phase chosen is nonpolar. On the other hand, a mixture of low molecular weight alcohols is better separated if a more polar and hydrogen-bonding phase is employed.

Gas chromatographic separations depend on the different rates at which the components of a sample travel through the column. If two substances, Y and Z, have the same affinity for the liquid phase, but Z is more volatile than Y (that is, has a lower boiling point), then Z will remain in the gaseous phase longer as it travels down the column than does Y. Thus, Y and Z are separated on the basis of their boiling points. In general, substances are separated on the basis of their boiling points if the polarities of the compounds are about the same. Table 17-1 lists the boiling points of simple alkanes and can be used to predict the order of elution in a gas chromatographic separation.

The column is enclosed in an oven used to maintain a constant temperature or for changing the temperature.

The Detector Several types of detector are used in gas chromatography. The simplest and one of the most widely used types is the thermal conductivity detector (TCD). This type of detector uses a Wheatstone bridge circuit to detect the sample components as they exit to the detector. The carrier gas most often used in gas chromatography is helium, although nitrogen can also be used. Helium is preferred as it has a very high thermal conductivity that allows more sensitive detection of the components. Organic compounds have smaller thermal conductivities than helium, and when a compound elutes at the detector, the temperature of the detector filament increases. This sensor is one arm of a Wheatstone bridge circuit. The final result is an imbalance in the bridge that causes a flow of electricity to a recorder that responds to the current (or voltage).

Results of a Gas Chromatographic Analysis There are two important results of a chromatographic analysis: the time it takes for a particular compound to elute to the detector, the **retention time**, and the **peak area.** The retention time, R_t is a function of the column temperature, the flow rate of the carrier gas, the column length, the nature of the stationary phase, and the physical properties of the analyte. Generally, the chromatographic identification of a compound is made by determining the retention times for a series of standards (usually in a mixture, such as a series of *n*-alkanes) and then measuring the retention times for each compound in the unknown mixture. Then if the unknown mixture gives peaks that have identical retention times as the standards, it is assumed that those peaks are due to the same compound as in the known mixture. Unfortunately, just because two peaks have the same retention time does not guarantee that they are the same compound. Verification is frequently required, and can be accomplished by using a mass spectrometer coupled to the gas chromatograph (called a GC/MS). This technique is widely used for environmental analysis, drug testing and many other areas.

The peak area of a component of a mixture is proportional to the mass of that component in the mixture. Thus, the peak area can be used to calculate the relative amount of that component in the mixture. If peaks are sharp, they are almost triangular in shape and the area can be calculated using the product of one-half the base length times the height of the triangle. The total area is calculated and the percentage of any one peak component is its area times 100 divided by the total area.

Table 17-1 Boiling Points of Some Simple Hydrocarbons

Name	Boiling Point (°C)	Name	Boiling Point (°C)
Butane	-0.5	2,2,3-Trimethylpentane	109.8
Pentane	36	2,3,3-Trimethylpentane	114.8
Hexane	68	2,2,4-Trimethylpentane	99.2
Heptane	98.4	2-Methylheptane	118
Octane	125.6	3-Methylheptane	115–118
Nonane	150.8	4-Methylheptane	118
Decane	174.1	2,3-Dimethylheptane	141
Dodecane	216.3	2,4-Dimethylheptane	133
2-Methylbutane	27.9	2,5-Dimethylheptane	136
2,2-Dimethylbutane	49.7	2,6-Dimethylheptane	135
2,3-Dimethylbutane	58	3,3-Dimethylheptane	137
2-Methylpentane	60.3	4-Ethylheptane	141
3-Methylpentane	63.3	2-Methyloctane	142.8
2,2-Dimethylpentane	79.2	3-Methyloctane	144-145
2,3-Dimethylpentane	89.8	4-Methyloctane	142.4
3,3-Dimethylpentane	138.2	2,7-Dimethyloctane	159.6
2-Methylhexane	90	Cyclohexane	81
3-Methylhexane	92	Benzene	80.1
2,2-Dimethylhexane	107	Toluene	110.6
2,3-Dimethylhexane	116	*o*-Xylene	144
2,4-Dimethylhexane	111	*m*-Xylene	139
2,5-Dimethylhexane	109	*p*-Xylene	138

Source: R. C. Weast, Editor-in-Chief, *CRC Handbook of Chemistry and Physics*, 59th ed., CRC Press, Boca Raton, FL, 1979.

Pattern Recognition—Figure 17-2 (top) shows the gas chromatogram of a sample of number 2 fuel oil taken from a ship which ran aground near the mouth of the St. Johns River in Jacksonville, Florida. Figure 17-2 (bottom) shows the gas chromatogram of a sample from a tar ball, found on a beach a few miles from the ship that was extracted for hydrocarbons. Even a cursory examination indicates that the two samples came from the same source. This identification by pattern similarity is called "pattern recognition" and has been used for many years to identify environmental PCBs (they have a very characteristic chromatographic pattern).

Although such identifications can be put on a more quantitative basis, pattern recognition is very useful in environmental studies. It can be made more exact by acquiring an extensive series of gas chromatograms for many different sample types, such as a series of petroleum-based products, and then comparing a sample suspected of being one of these types with the "library," which was assembled from chromatograms of known sample types.

Figure 17-2 Gas Chromatograms of (top) Number 2 Fuel Oil Taken from a Ship that ran Aground off Jacksonville, Florida, and (bottom) Hydrocarbons from a Tar Ball Found Several Miles from the Oil Spill

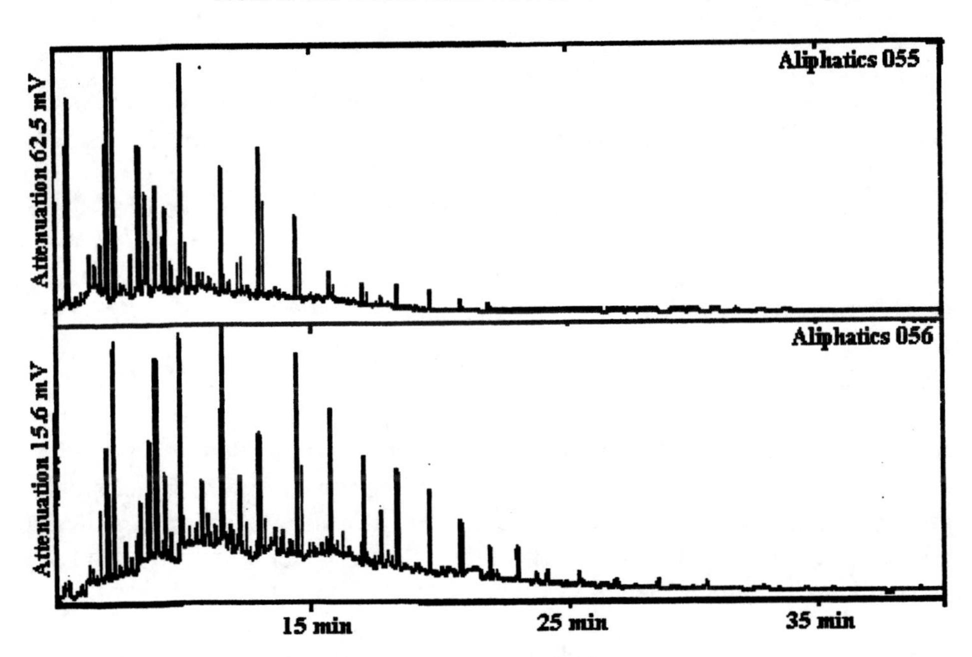

In this experiment, you will prepare a "mini-library" of reference chromatograms. These will be the chromatograms of the headspace vapors of different petroleum-derived sample types. Then the chromatogram of the unknown will be determined and by visually comparing the chromatograms, the nature of the unknown can be determined.

You will also determine the chromatogram of a series of simple alkanes and another of low molecular weight alcohols. Because retention times are linearly related to their boiling points, the retention times will be characteristic of each compound. Then by comparing retention times of the standards with those in the petroleum-derived samples, some of the compounds in the unknowns can be identified.

One potentially confusing point is that standard alkanes and alcohols will be injected in liquid form in the microliter volume range. The sand (or soil) samples will be examined for compounds in the vapor phase at room temperature. Since liquids are about 1000 times more dense than the vapor, a sample of at least 1 mL of vapor must be injected to obtain a comparable signal. Fortunately, the sensitivity range of the thermal conductivity detector is wide enough to sense small concentrations. Based on comparing the R_t values of the standards with the unknowns, you will get a more quantitative "feel" for the identity of your contaminant.

Safety Precautions

1. Safety glasses must be worn at all times in the chemistry laboratory.

2. Do not stand in front of a reduction gauge on a gas cylinder when opening the cylinder. There is a danger of its rupturing when opening the main valve.

3. The injection port of the gas chromatograph will be very hot. Be careful not to touch it.

Procedure—A mixture of *n*-alkanes, from pentane through nonane, and one of alcohols, from methanol through butanol, are used to obtain chromatograms. The mixtures have been prepared using different, and known, quantities of each component. One objective of this experiment is for you to determine the percentage of each component in each mixture using peak areas from the chromatograms.

Second, the headspace vapors from each reference petroleum-derived mixture and the unknown, will be injected into the gas chromatograph and their gas chromatograms obtained.

Injection of the Liquid Samples

Table 17-2 gives suggested injection volumes, but the actual volume injected depends on the sensitivity of your chromatographic system. To avoid overloading the column, it is best to inject small volumes and increase the sample size if necessary. It saves time if several students work together to obtain the chromatograms and then share copies. However, each student should inject at least three samples. Inject the following samples and obtain the chromatograms.

1. Inject 2–3 µL of the alcohol mixture and obtain the chromatogram. Thoroughly rinse the syringe with pentane by drawing pentane into the syringe and plunging it out into a waste container. This should be repeated at least five times to thoroughly clean the syringe.

2. Repeat Step 1 using the mixture of *n*-alkanes.

The conditions for a typical chromatographic analysis of alkanes are given in Table 17-2.

Table 17-2 Chromatographic Conditions for the Separation of Alkanes

Chromatographic Variables	Typical Value(s)
Detector current	150 mA
Column(Choice)	4' length x ¼ in. diameter; 20% Carbowax
	20M or DC200 on Chrom P(80–100 Mesh)
Column temperature	110°C
Flow rate	40–45 mL/min
Recorder scale	10 mV
Chart speed	2.5 cm/min
Attenuation	2(or as needed)
Injection volume, liquid	2–3 μL
Injection volume, vapor	1 mL

Headspace Analysis

1. Erlenmeyer flasks containing soil or sand have been treated with samples of gasoline, kerosene, jet fuel, shellac thinner, and mineral spirits. Each flask contains a different sample, and the tops of the flasks are sealed with septa. Do not remove the septa as volatile components in the gaseous phase may be lost, which will alter the composition of the head space gases.

2. Each student selects a known sample, which is identified, and withdraws a 1 mL sample of the headspace gases. This sample is then injected into the gas chromatograph. The syringe must be gas-tight. The gas chromatogram is then obtained. The collection of chromatograms for the known samples is the "mini-library." This procedure is repeated until all the known samples have been injected.

3. Each student is then given an unknown sample to identify. The chromatogram is obtained in the same way as for the known samples.

Waste Minimization and Disposal

1. If the normal alkanes and alcohols are in containers with rubber septa, they are stable indefinitely and do not have to be disposed of. If the alkane and alcohol mixtures need to be disposed of, they are turned over to a commercial hazardous waste contractor.

2. The samples of gasoline, kerosene, etc. in sand or soil are hazardous wastes and can be combined and disposed of by a commercial facility. Alternatively, the samples can be used for a period of years if additional sample is added to the matrix.

Data Analysis

1. For the standard alcohol and alkane mixtures, write the retention times above each peak on the gas chromatogram. If the peaks are triangular, estimate the areas by multiplying one-half the base length times its height. If peaks are not sharp, estimate the areas by counting the number of squares under each peak on the chart paper. The percentage composition of a particular component is its area divided by the total area times 100. Calculate the percentage composition of each component in each mixture.

2. Prepare a table of retention times for each component and its percentage in the mixture.

3. Prepare two plots, one for each standard mixture, of retention time (ordinate) versus the boiling point of the compound. Carry out a least-squares analysis and obtain the correlation coefficient.

4. Describe the pattern of chromatographic peaks for the unknown sample for which a headspace analysis was performed. Which pattern for the known samples most closely resembles this pattern? The early peaks are often the most relevant for this comparison since they have the most reproducible retention times.

5. Gas chromatograms for liquid samples of gasoline, kerosene, etc., will be available for comparisons.

Supplemental Activity

1. A wide variety of petroleum-derived products can be examined in this experiment. Samples of high volatility are preferable since higher-boiling components will not elute from the column.

2. To have more time with the gas chromatograph, one of the more important tools in environmental science, the student can be allowed to determine the effect on the chromatogram of the *n*-alkanes by changing the following: flow rate, column temperature, and sample size.

3. It would be informative to warm the headspace samples by about 10°C and obtain a new chromatogram to see the effect of the increased temperature.

Questions and Further Thoughts

1. Which mixture separates better (i.e., has the greater distance between peaks), the alcohols or the alkanes? Explain the difference.

2. Which alcohols are present in shellac thinner? Are alcohols present in any of the other samples?

3. Which alkanes are present in gasoline, and which are present in the other samples?

4. Use the data of Table 17-1 to plot boiling point (ordinate) versus carbon number for the *n*-alkanes. Discuss the linearity with respect to the intermolecular forces. Is there a relationship between R_t and the boiling points? Make tentative boiling-point assignments for the peaks observed in the headspace of your soil sample. Assuming it is a hydrocarbon, give a tentative assignment for each peak.

Notes

1. Many terms are important in the field of gas chromatography, such as solid phase, absorbent, eluent, relative retention time, support, etc. It may be helpful for the student to compile a "mini-dictionary" of these "buzzwords," using several literature sources.

2. It is important that the gas chromatograph be ready when students begin this experiment. It takes over an hour for the thermal conductivity (TC) detector to stabilize.

3. When the detector current is on there must be a flow of carrier gas through the detector. Otherwise, the detector will be destroyed.

Literature Cited

1. A. P. Bentz, Oil spill identification, *Anal. Chem.48*(6):454A–472A (1976).

2. D. L. Duewer, B. R. Kowalski, and T. F. Schatzki, Source identification of oil spills by pattern recognition analysis of natural elemental composition, *Anal. Chem.,47*(9):1573–1583 (1975).

3. J. S. Fritz and G. H. Schenk, *Quantitative Analytical Chemistry,* 5th ed., Allyn and Bacon, Boston, 1987.

4. R. C. Weast, Editor-in-Chief, *CRC Handbook of Chemistry and Physics,* 59th ed., CRC Press, Boca Raton, FL, 1979.

Experiment 18

Analysis of Environmental Hydrocarbons Using Simple Extraction and Analysis by Flame-Ionization Detection Gas Chromatography

Objective—The objective of this experiment is to introduce you to one of the most important analysis methods used in environmental chemistry, flame-ionization detection gas chromatography (FID/GC). You will also learn how to extract trace amounts of hydrocarbons as well as the higher levels that would be found in an oil spill.

Introduction— Hydrocarbons are particularly important pollutants because of the large volumes of petroleum transported, spillages in transfer, and the impact from sewage outfalls and stormwater runoff. Small quantities of oils and greases can contaminate large quantities of water, and the aromatic components can be harmful to shellfish and finfish. Many aromatic compounds and their metabolites are potent carcinogens.

Petroleum, or crude oil, is a mixture of hundreds of compounds, most of which are hydrocarbons (HCs). Hydrocarbons are compounds of carbon and hydrogen and are structurally classified as aliphatic or aromatic. The aliphatic HCs may be straight-chain (called normal) or branched. Aliphatic HCs include the alkanes and cycloalkanes, which are illustrated in Figure 18-1. Aromatic HCs are characterized by a ring structure with alternate single and double bonds as shown in Figure 18-2. Some monocyclic aromatic HCs associated with pollution from gasoline include benzene, toluene, ethylbenzene, and the xylenes.

Fused ring systems, shown in Figure 18-3, are called polycyclic aromatic hydrocarbons (PAHs) or polynuclear aromatic hydrocarbons (PNAs), some of which are the most carcinogenic substances known.

Hydrocarbons found in the environment are mainly of petrogenic origin. However, low levels are ubiquitous in nature. Elevated levels are usually associated with human activities and thus are termed anthropogenic hydrocarbons. This term usually refers to refined products obtained from petroleum.

Petroleum refining begins with the separation of crude oil into various fractions by fractional distillation. Several important fractions are shown in Table 18-1. The origin of HCs in the environment is given in Table 18-2.

The Effects of Environmental Petroleum

Fish exposed to petroleum in water, sediments, and food supply rapidly take up HCs which accumulate in tissues of the liver, brain, and muscle.

The predominant chemical change in petroleum on entering the environment is oxidation. Oxidation occurs either through photochemical reactions or enzymatic reactions in microorganisms. Hydrocarbons that dissolve in water quickly evaporate or become diluted and are ultimately flushed from the stream.

Since petroleum HCs are hydrophobic, and thus lipid soluble, a major transport mechanism is via their association with suspended particulates. Sediments are major reservoirs (sinks) for such pollutants.

Microbial degradation is the most important process involved in the weathering of petroleum in the environment, and the rate decreases in the following order: *n*-alkanes > branched alkanes > aromatics > cycloalkanes.

Figure18-1 Structures of Aliphatic Hydrocarbons

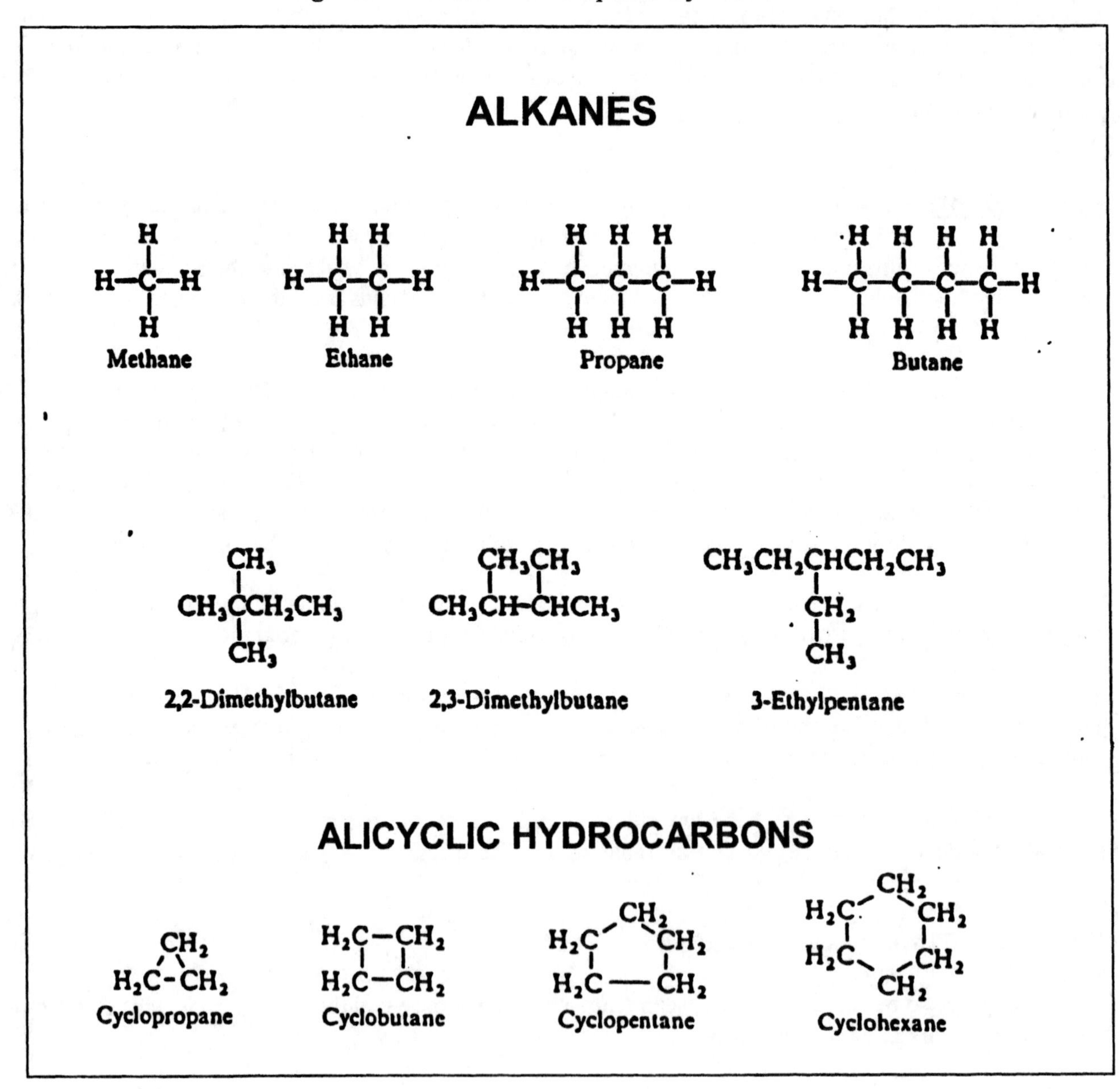

Figure 18-2 The Simplest Aromatic System, Benzene

Figure 18-3 Polycyclic Aromatic Hydrocarbons

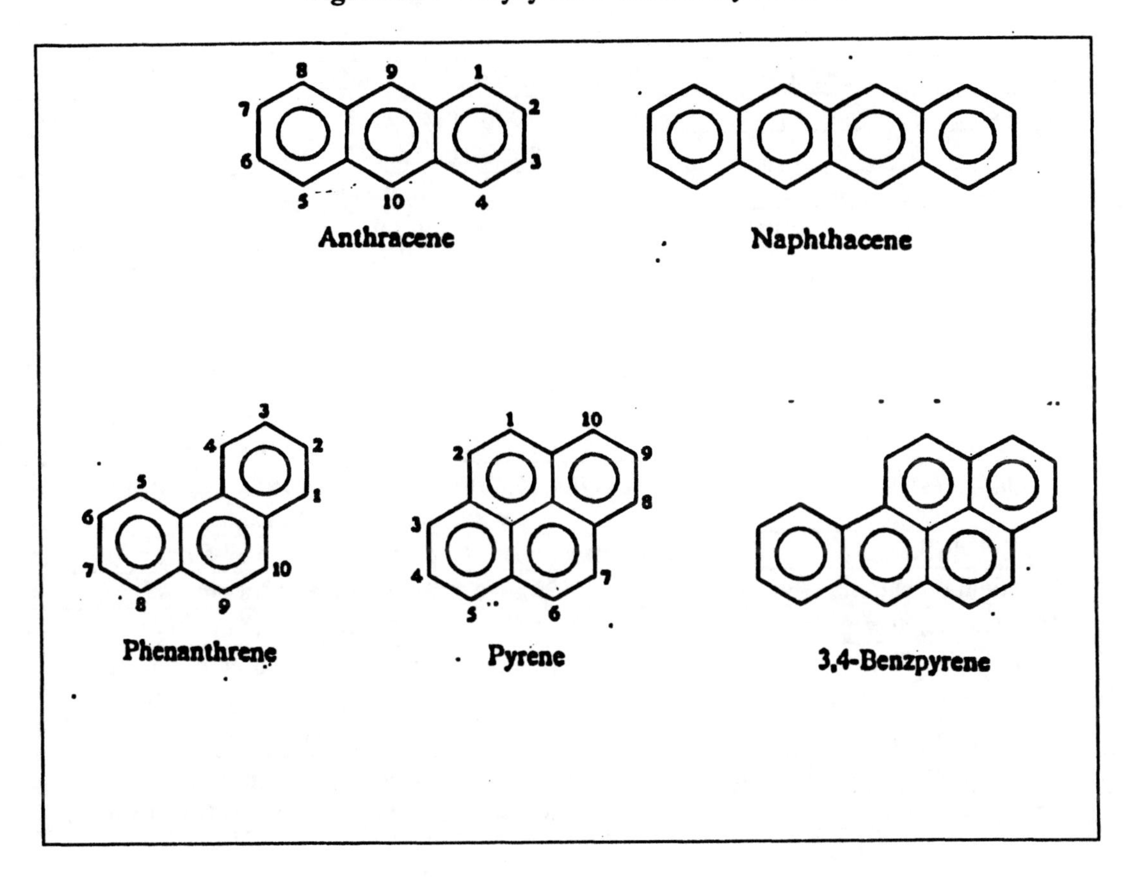

Table 18-1 Substances Obtained by the Refining of Petroleum

Carbon Content	Name of Fraction
C_1–C_4	Natural gas
C_5–C_6	Petroleum ether
C_6–C_7	Ligroin (light naphtha)
C_6–C_{12}	Gasoline
C_{12}–C_{18}	Kerosene
$> C_{18}$	Gas Oil (furnace oil, diesel oils)
$> C_{18}$	Lubricating oils
$> C_{18}$	Waxes
$> C_{18}$	Asphalt

Fuel Oils

Number 1	Kerosene, range oil, aviation and diesel fuels, domestic heating
Number 2	Domestic heating
Number 4	Light industrial fuels
Number 5	Industrial burners (needs preheating)
Number 6	Ships, industry, large-scale heating (needs preheating)

Table 18-2 How Petroleum Enters the Environment

Tankers	Oil-water mixtures remaining in tanks are Discharged into the environment
Tanker accidents	The major cause is structural failure, followed By grounding and collisions
Dry docking	Tankers must be clean during maintenance and inspection
Terminal operations	Losses from spillage in transfer
Bilges and bunkering	Bilge water and leaks from bunkers
The atmosphere	Internal combustion engines and power plants
Municipal and industrial wastes	Sewage-treatment plants and industrial wastewater
Urban runoff	Oil-heating systems, automobiles, and service stations
Point sources	Power plants, military bases, and marinas
Storage facilities	Gasoline and jet fuel tanks, military fuel depots

Theory—The HCs most readily identifiable by gas chromatography are the normal alkanes. These HCs are separated on the basis of their boiling points, the lower boiling point compounds elute from the column first and the least volatile ones elute at a later time. Figure 18-4a shows the gas chromatogram of an environmental sample that illustrates the spacings of the *n*-alkane peaks from a relatively clean environment. Figure 18-4b shows the chromatogram of the HCs from a contaminated, but unweathered, environment. The hump is characteristic of petrogenic HCs. Figure 18-4c shows the chromatogram of a weathered environmental sample of petrogenic HCs. The normal alkanes have been largely biodegraded, leaving unresolved cycloalkanes in the hump.

Figure 18-4a Gas Chromatogram of a Relatively Clean Environmental Sample Showing the Peak Spacings of the *n*-Alkanes

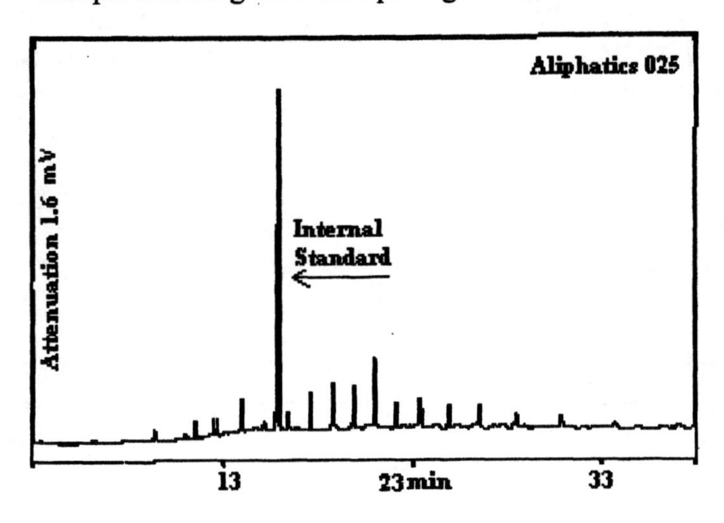

Figure 18-4b Gas Chromatogram of the Alkanes from a Contaminated Unweathered Sample of Petrogenic Hydrocarbons

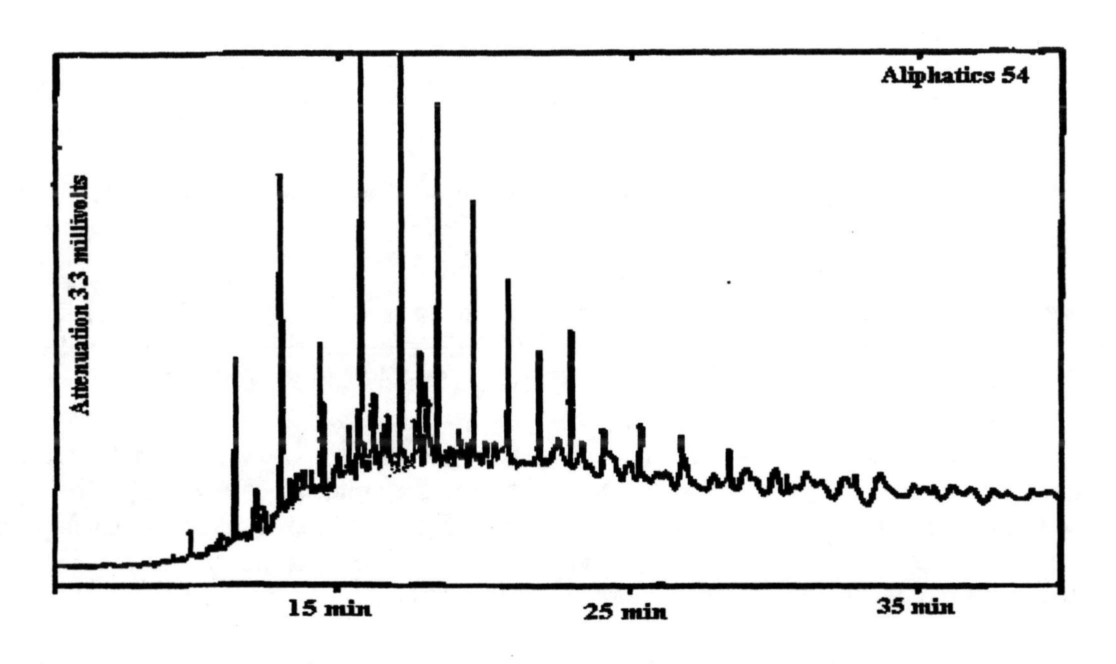

Recently contaminated sediment samples often result in gas chromatograms with alkane peaks ranging from fewer than 10 to about 30 carbon atoms. Generally, however, alkanes with fewer than about 12 carbon atoms are too volatile to persist in the environment for any length of time. Being hydrophobic, if there is no suspended silt onto which the HCs can sorb, they will tend to migrate to the surface where hydrophobic interactions are less, and at the surface they can rapidly evaporate.

Due to the great range of alkanes usually present, temperature programming is necessary to decrease retention times for the larger alkanes. This is also necessary to keep peaks sharp. In temperature programming the temperature is increased at a certain rate (experimentally determined) until it reaches a maximum value, which is often held for a period of time to allow the less volatile components to elute.

Since isolating HCs is best accomplished by extraction with hexane, hexane predominates in the extract and would tail (that is, overlap other peaks) excessively on the gas chromatogram and thus interfere with the quantitation of even the heavy alkanes. To avoid this problem, a septum purge is done after injection, eliminating most of the tailing. This technique eliminates excess solvent from the region near the septum and thus the excess solvent never enters the column.

Figure 18-4c Gas Chromatogram of a Weathered Sample of Petrogenic Hydrocarbons

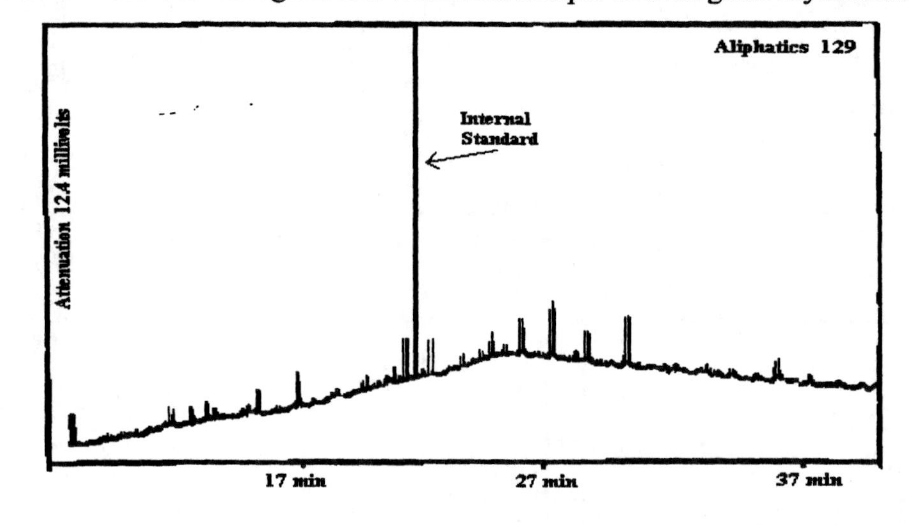

Because HCs are flammable, flame-ionization detection is used to detect HCs emerging from the column. A flame ionization detector (FID) is at least 1000 times more sensitive than a thermal conductivity detector for typical HCs. The diagram of a typical FID system is shown in Figure 18-5.

In a FID a very hot flame is produced by burning hydrogen in air. The high temperature causes the production of ions as an eluting compound is burned. The ions are collected at a charged electrode and the resulting current is measured with an electrometer amplifier. The burner jet is made the negative electrode, and a loop of inert metal surrounding the flame is made positive. The sensitivity of the FID to organic species is about proportional to the number of carbon atoms present.

Quantitation of the gas chromatogram can be done manually using peak heights, but it is better to use a computerized data-handling system, preferably one that can be used to calculate results and change the attenuation after the chromatographic run.

Figure 18-5 The Flame Ionization Detector

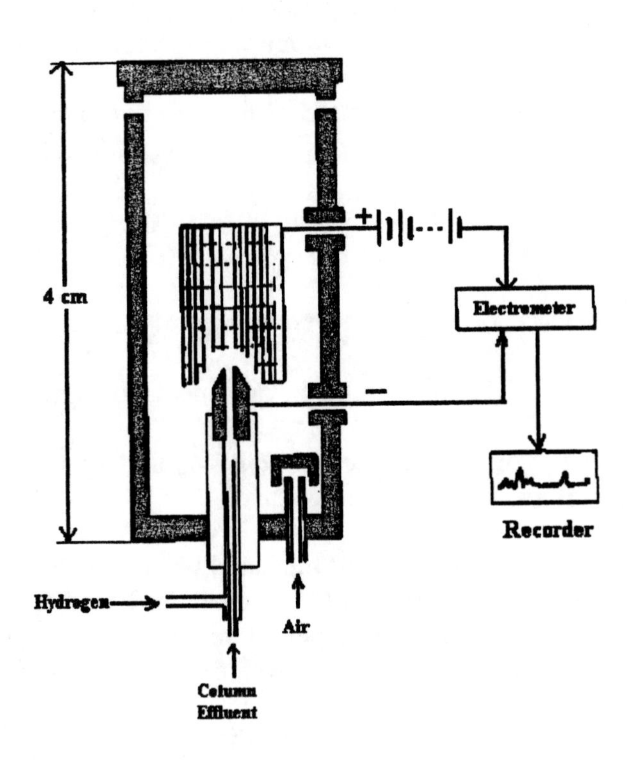

Source: G. W. Ewing, *Instrumental Methods of Chemical Analysis*, 5th ed., McGraw-Hill, Inc., New York, 1985, p. 360. By permission.

Safety Issues

1. Safety glasses must be worn during all parts of this experiment.
2. The instructor must be present when the gas chromatograph is being used.
3. Use gloves when handling aromatic substances.
4. Avoid physical contact with toluene and also avoid breathing its vapor.

Procedure

A: Sample Collection and Storage

Avoid using plastic devices for sample collection or storage. Plastics contain plasticizers that can interfere with the analysis. In fact, if sediment samples are collected from an urban stream, plasticizers will frequently be extracted with the HCs, and they often dominate the gas chromatogram.

Water samples are not easily studied for HCs since HCs are only slightly soluble in water. An area where there has been an oil spill that has washed up onto a sandy beach is ideal for sample collection. Tar ball samples remain unchanged in composition for a long time. An area where there has been considerable fuel transportation or storage is also good for sampling. A marina area also provides excellent sediment samples for analysis. To minimize the amount of material that will have to be disposed of eventually, collect a minimum amount of sample. About 100 g per group is appropriate for this experiment.

Since many bacteria metabolize HCs, samples should be extracted soon after collection. Otherwise, they must be stored in a refrigerator at 4°C.

B: Sample Extraction and Preparation for Gas Chromatography Analysis

If the sample has a **substantial HC level,** a simple hexane/methanol extraction will suffice.

1. Excess water is first decanted from the sediment (or soil) sample. Then a sample with mass of about 10 g (weighed to 0.1 mg) is weighed into a tared 150 mL beaker.

2. A 50 mL solution of pesticide grades hexane and methanol (4:1) is added to the beaker followed by 0.25 mL of the internal standard (I.S.), added with a pipet. A small stir bar is added and the beaker is put on a magnetic stir plate in a fume hood. A watchglass is placed on the beaker and the sample is stirred for about 0.5 hour. For a sandy matrix less time will be required, whereas more time is needed for a clay-like matrix.

3. After extraction, the sample is gravity filtered (medium porosity paper, Whatman Number 1, for example) into a 150 mL beaker. The extraction beaker and funnel are rinsed with two 5 mL portions of hexane.

4. The combined filtrate and rinsings are dried with anhydrous sodium sulfate, using a magnetic stir bar to stir. The sample is then filtered into a 150 mL beaker, rinsing the previous beaker and funnel with two 5 mL portions of hexane.

5. At this point, nitrogen gas could be used to evaporate excess solvent, or a concentrator, such as a Kuderna-Danish, could be used. However, it is convenient and much less expensive to evaporate the solvent in a fume hood for 6–12 hours. An aluminum tent over the beaker prevents contamination.

6. The sides of the beaker are occasionally rinsed down with hexane, and the evaporation is continued until the final volume is 1 to 2 mL. The sample is now ready for gas chromatography.

An alternative extraction procedure uses the Soxhlet apparatus described in Experiment 14. The Soxhlet procedure is important for the extraction of **very low levels of HCs** but requires much more time and equipment.

C: The Internal Standard

The I.S. solution is prepared by weighing out 10 mg of androstane in a small beaker, dissolving in hexane, and then quantitatively transferring to a 100 mL volumetric flask. The solution is diluted to the mark with hexane and then thoroughly stirred.

If the final volume of the extract before chromatography is about 2 mL, the I.S. concentration will be about 10 ppm if the I.S. volume used is 0.25 mL. Relatively clean environmental sediments have a total HC level less than 10 ppm, whereas very contaminated samples may have total HC levels several orders of magnitude greater. The I.S. may be provided.

D: Gas Chromatography Analysis

Any gas chromatographic system that provides an FID, temperature programming, and computerized data handling can be used. The following instructions apply to a Perkin-Elmer Sigma 3B gas chromatograph with a microprocessor to adjust injector and detector temperatures as well as carry out temperature programming. Instructions will be similar for other instruments.

A DB-5 fused silica capillary column, 0.25 mm in diameter and 30 m in length, gives excellent peak resolution. A constant split of about 10 mL/min is used to maintain a steady flow of gas away from the septum area to prevent back diffusion. Thirty seconds after injection, the septum purge valve is opened to sweep away solvent that continues to desorb. The purge prevents severe tailing of the solvent peak. A purge rate of 4–5 mL/min is used.

An analysis consists of a 2.0–3.0 μL sample injection, the initial temperature of 100°C held for 0 minutes. This is followed by an 8°/min temperature ramp to a final temperature of 280°C. If only lighter HCs are present, this temperature needs to be held for only 5 minutes, followed by automatic cooling back to 100°C.

1. A 10 μL syringe is rinsed at least five times with pesticide grade hexane. It is then rinsed at least two times with the alkane standards. The rinsings are put into a waste vial that is provided.

2. The syringe is filled to 2–3 μL with the standards, any excess sample being injected into the waste vial. The sample is then injected into the gas chromatograph quickly and smoothly. The start button on the instrument is quickly pressed. After 30 seconds the septum purge is opened. **This is a critical step** as it prevents tailing of the hexane peak!

3. The instrument will now produce a gas chromatogram and a report giving the retention times and areas of each peak and additional results if it has been instructed to do so.

4. The syringe is rinsed at least 10 times with hexane.

5. The syringe is then rinsed twice with the environmental HC extract, and then filled with 2–3 μL of extract.

6. The sample is injected into the gas chromatograph, as before. Again open the septum purge after 30 seconds. (The exact time is not critical, but should be between 25 and 35 seconds.)

7. The instrument and data station generate a chromatogram and report as before, but there should be many more peaks on the chromatogram than the previous sample.

E: Data Handling

The gas chromatogram is quantitated by first determining the area corresponding to the I.S. The concentrations of the *n*-alkanes can be calculated by using their areas and assuming a response factor (R.F.) of unity for each. For equal masses the response factor compares the area of a component with the area of a reference peak, such as the internal standard, which is defined as having a response factor of unity. A response factor of 0.85 for a component means that its area is 0.85 times the area for the internal standard if the masses of both are equal in the mixture.

For a HC analysis, an accurate integration of peak areas is required and it is necessary to have the capability of handling a large number of peaks. A Perkin-Elmer Sigma 3600 Data Station is one of several systems that fills the requirements. This device gives the area for each peak and by using the I.S. concentration and the original sample mass, it converts areas into concentrations.

The standard alkane mixture is chromatographed to determine retention times for individual alkanes. The alkanes from dodecane to eicosane will be present in the standard mixture. If time is a factor, a copy of the chromatogram for the standard alkane mixture will be provided.

An option in this experiment is to use the known masses of the alkanes in the standard mixture, and the corresponding areas to calculate response factors for the alkanes. However, these are often close to unity. It is also straightforward to calculate the percentage composition of each alkane and compare the results with the given composition.

After the standard alkane chromatogram has been obtained, the retention time for each component can be entered into the data station software so that when the environmental sample is run, the data station will identify the alkanes and report their areas and the ppm of each present in the mixture.

F: Optional Separation of Aliphatics and Aromatics

An option for this experiment is to separate the aromatic and aliphatic portions using gravity-column chromatography. Since usually only small amounts of HCs are present, a very small column can be used for the separation and the experimental time is short.

1. Hexane is used to add a slurry of silica gel to fill a small column (5 cm or less) to about 40% of its capacity. This is followed with a hexane slurry of alumina, filling an additional 40 percent of the column. A thin layer of clean sand is placed on top of the alumina.

2. The sample is placed at the top of the column and is eluted with three column volumes of hexane. The hexane is then replaced with toluene and the column is eluted with three column volumes of eluent. Evaporation, as previously described, to 2 mL is then followed by gas chromatography. In this option, *ortho*-terphenyl is used as the I.S. for the aromatic fraction. (Use 0.10 mL of 10 mg *ortho*-terphenyl dissolved in 100 mL of hexane, added before the initial extraction with hexane.)

3. If a sample is particularly heavy in HCs, it is possible to quantitate the masses of the two fractions by completely evaporating the solvents. Otherwise, they can be quantitated by using the I.S.

Waste Minimization and Disposal

1. Since interesting samples are contaminated with petroleum hydrocarbons, collect only a minimum amount of sediment or soil since extra material must be disposed of as hazardous waste. The extracted samples must also be treated as hazardous wastes as well and are transferred to a bottle for storage. This will later be given to a hazardous waste contractor.

2. The *n*-alkane standards and the two internal standards, androstane and *ortho*-terphenyl, should be kept in containers with rubber septa where they can be kept for an indefinite period of time, and used in future experiments.

3. The alumina and silica gel (if used) and sodium sulfate are slightly contaminated with hydrocarbons and must be considered to be hazardous wastes. These three substances can be combined in a bottle for later disposal by a waste contractor.

4. Any waste hexane and methanol must be stored in separate bottles for later commercial disposal. Waste toluene should be kept in a separate bottle for later commercial disposal.

Data Analysis

1. Give full sampling details.

2. Submit your chromatogram(s) and the report(s) generated by the data station. Describe the pattern of aliphatic peaks and whether or not there is a hump in the chromatogram. Try to decide if the HCs in the sample are fresh or weathered.

3. Calculate the percentage of each alkane in the standards mixture. Compare with the given results that come with the standards. (The standards may be purchased or may be made up by the lab instructor.)

4. Use the report generated by the data station for the environmental sample to obtain the sum of the ppm of the alkanes of odd numbers of carbon atoms and do the same for the alkanes with an even number of carbon atoms. The carbon preference index (C.P.I.), is the ratio of odd to even hydrocarbons. Find this ratio for your sample. A large C.P.I indicates HCs of biogenic origin. Petrogenic HCs, on the other hand show little preference for either odd or even HCs and the C.P.I. is closer to unity. Also, a longer *n*-alkane range is found with petrogenic HCs; natural HCs often have a narrow range of *n*-alkanes.

5. Report the total HC area and the area for the I.S. Use the amount of I.S. and its area to calculate the total ppm of HCs present in the sample.

6. If the HCs were separated into aliphatic and aromatic fractions, determine the ratio of the two, gravimetrically (if this was done) and by using the internal standards to quantitate the two fractions (if this was done).

7. Discuss the extent of HC contamination of the sample.

Supplemental Activity

1. Instead of environmental samples, samples of petroleum or petroleum products can be studied. Samples of petroleum from different parts of the United States give very different gas chromatograms as do samples from different parts of the world.

2. If aliphatics and aromatics were separated, the aromatics fraction can be examined by fluorimetry as in Experiment 15.

Questions and Further Thoughts

1. Benzene has been used to separate aliphatics and aromatics, but this practice is discouraged because benzene exposure is known to cause leukemia. The disadvantage of toluene is its lower volatility.

2. Some environmental samples have aliphatic hydrocarbons with more than 30 carbon atoms. In this case the hold time at the higher temperature must be increased to purge these heavier hydrocarbons from the column. If this is not done, these compounds will elute with the next injected sample. If broad peaks appear at the start of a new chromatographic run, this is a likely source of the peaks.

Notes

1. Do not exceed the maximum temperature recommended for the column. Higher temperatures can cause the stationary phase to elute or decompose, thus ruining the column—a costly mistake!

2. A new column must be conditioned by allowing carrier gas to pass through it at an elevated temperature (above the temperature at which it will be used) for several hours. When conditioning the column, it must not be connected to the detector or the detector may be harmed!

Literature Cited

1. D. N. Boehnke, C. A. Boehnke, K. I. Miller, J. S. Robertson, and A.Q. White, *Final Report, Survey of Hydrocarbons in the Lower St. Johns River in Jacksonville, Florida*, prepared for the Florida Department of Natural Resources, Tallahassee, FL, October 1983.

2. M. Grayson, Executive Ed., *Kirk-Othmer Encyclopedia of Chemical Technology*, Wiley Interscience, Ney York, 1982: Vol. 11, pp. 317-333, 652-695; Vol. 17, pp. 110-131, 183-190, 199-256, 257-271.

3. S. E. Hamilton, T. S. Bates, and J. D. Cline, Sources and transport of hydrocarbons in the Green-Duwamish River, Washington, *Environ. Sci. Technol. 18*(2): 72–79, (1984).

4. D. MacLeod, G. Prohaska, D. G. Gennero, and D. W. Brown, Interlaboratory comparisons of selected trace hydrocarbons from marine sediments, *Anal.Chem., 54*:389–392 (1982).

5. J. E. Ravan, *Regional Coastal Pollution Control,* B. L. Edge, ed., Coastal Zone Pollution Management, Proc. of the Symposium, Clemson University, Clemson, SC, 1972.

6. E. S. Van Vleet and J. G. Quinn, Input and fate of petroleum hydrocarbons entering the Providence River and Upper Narragansett Bay from wastewater effluents, *Environ. Sci. Technol., 11(12):1086–1092,*(1977).

Experiment 19

Kinetics of the Decomposition of Pollutants in the Environment with an Application to Plasticizers

Objectives—The objectives of this experiment are:

1. To introduce you to some basic concepts of chemical kinetics that are important to the environment

2. To show how an indirect method, the use of solvent mixtures, can get around the problem of low solubility and slow rates of environmental reactions

3. To illustrate the kinetics of an important reaction type, saponification

4. To introduce you to plasticizers, which have become almost ubiquitous in the environment

Introduction—When pollutants enter the environment, the first question we ask is, where do they go? followed by, how long will they remain there? Many chemical reactions occur in the environment, and the rate at which they occur is of primary importance. We would like to know, for example, how long DDT will continue to impact agricultural areas where it was once used.

Certain substances, upon entering the environment, "disappear" quickly, being converted into other chemical substances. Hydrochloric acid, when flushed down a drain in a chemistry laboratory, will be quickly neutralized by basic substances, including carbonates and bicarbonates. Polyvinyl chloride waste, from a manufacturer of plastics, on the other hand, may persist in the environment for a long time.

Whether a chemical reaction is favored or not lies in the realm of chemical thermodynamics; on the other hand, studies of the rates of reactions lie in the province of chemical kinetics. A reaction may be thermodynamically favorable but so slow that its significance is minimal.

Pollutants in the environment are exposed to a number of factors that can alter them, including water, oxygen (in the atmosphere and dissolved in water), sunlight, and bacteria. The effectiveness of the environmental factors varies greatly with the type of pollutant and its physical and chemical characteristics. A low molecular weight alcohol, for example, will be volatile and will be subjected to solar radiation in the atmosphere as well as oxygen. A higher molecular weight alcohol, on the other hand, is less volatile and yet still has significant solubility in water. It thus will be subject to dissolved oxygen in water and aerobic bacteria. If the substance is of even greater molecular weight, its water solubility will become negligible and molecules of the substance will become attached to organic matter in soil and water, eventually settling out into the sediments of rivers and streams.

One of the most important environmental chemical reactions is hydrolysis. Although hydrolysis can occur in the atmosphere, it is more likely to be important in the hydrosphere where the water concentration is much greater (55.6 M).

Reactions of substances with water are called hydrolysis reactions. Important pollutants that react with water are often organic and include pesticides, PCBs, plasticizers, detergents, and many others. The classes of organic compounds important in hydrolysis reactions include esters, amides, and halogenated hydrocarbons. In this experiment we shall examine the hydrolysis of a plasticizer that has an ester as a functional group.

Theory—The general equation for the hydrolysis of a carboxylic ester is

$$\mathbf{RCOOR' + H_2O \longrightarrow RCOOH + R'OH}$$

where R is called the acyl substituent and R' the alkyl substituent. The reaction is called saponification, and the reverse reaction is esterification. The R and R' groups can be methyl-, ethyl-, etc., or aromatic, including the phenyl group and so on.

Saponification can occur in neutral solution but it is quite slow. It is sped up dramatically in the presence of either dilute acid or base. In the presence of base, the reaction is written,

$$\mathbf{RCOOR' + OH^- \longrightarrow RCOO^- + R'OH}$$

The rate law for this reaction is second-order overall, first-order in each of the reactants. The rate law may be written

$$\mathbf{Rate = \mathit{k}[ester][OH^-]} \qquad \mathbf{(19\text{-}1)}$$

where the brackets refer to molar concentrations. To simplify the experiment, and the mathematics, the initial concentrations of ester and base are made equal. This allows the following expression to be derived, which gives the concentration of ester (or base) at various times:

$$\mathbf{1/[ester]_{\mathit{t}} = 1/[ester]_o + \mathit{kt}} \qquad \mathbf{(19\text{-}2)}$$

where $[\text{ester}]_0$ is the initial ester concentration and $[\text{ester}]_t$ is its concentration at any later time t.

The rate constant, k, can be obtained by measuring the ester (or OH^-) concentration at various times and calculating and averaging values of k. Alternatively, a plot of $1/[\text{ester}]_t$ versus time is made to give a straight line, the slope of which is equal to k.

Ester or hydroxide concentrations can be determined several ways. If the ester absorbs in the UV or visible region of the spectrum, its concentration can be monitored by measuring the absorbance at various times (since the absorbance is proportional to concentration—Beer's law). Alternatively, monitoring the pH will give the hydroxide concentration at various times. Also, if the reaction is slow enough, aliquots of the reaction system can be taken and titrated with standard acid.

On examining the hydrolysis reaction note that OH^- is being replaced by the carboxylate anion. The OH^- ion is an excellent conductor of electricity (as seen in Table 6-2), whereas the carboxylate ion is less so. Thus, the reaction can be monitored by measuring the electrical resistance (or conductance) of the reacting system at various times.

The equation that relates conductance to time is

$$L_t = (1 / a_0 k)(L_0 - L_t) / t + L_\infty \qquad (19\text{-}3)$$

where L_t is the conductance at time t, L_0 is the initial conductance, L_∞ is the conductance when the reaction is complete, and a_0 is the initial concentration of either reactant (if they are equal; if not, their average concentration is used as long as they are not very different). If L_t is plotted as ordinate and $(L_0 - L_t)/t$ as abscissa, the straight line will have a slope of $1/a_0k$.

The saponification reaction has been studied in great detail. Substituents in both the acyl and alkyl parts of the molecule influence the reaction rate. A substituent group may influence the rate of hydrolysis through an inductive (electrical) or resonance effect or through a steric (size) effect. Inductive or resonance effects influence the charge distribution in the molecule. Steric effects involve the size of the substituent groups adjacent to the site where the reaction occurs.

Plasticizers

If fish or sediments from an urban stream are extracted for hydrocarbons, large amounts of unexpected substances are frequently found. An analysis by mass spectrometry shows that these substances are esters of carboxylic acids and the origins of such substances are plastics.

Many plastics in pure form are useless for any application. Polyvinyl chloride (PVC), for example, is a gummy material that was originally thought to be useless. However, certain chemical substances dramatically alter the properties of a pure polymer. A plasticizer is a substance that is added to a polymer to increase its desirable characteristics, properties such as flexibility and workability.

Organic plasticizers are frequently liquid esters of moderately high molecular weight. Hundreds of plasticizers are available but only about 100 are commonly used. Adipates are low temperature plasticizers that are commonly found in the environment. Phthalate esters, particularly dialkyl phthalates, have been widely used in plasticizer technology since the 1930s. One example of this type, bis(2-ethylhexyl)phthalate, was the accepted industry standard for many years,

Because of the ubiquity of plastics, plasticizers are widespread in the environment. Chemically, they can undergo hydrolysis to carboxylate ions and alcohols, enhancing their solubility in water and thus their potential for further degradation via oxidation, photolysis, further hydrolysis or biodegradation. Also, there are probably no plasticizers that are free from attack by fungi or bacteria.

Plasticizers are persistent in the environment for several reasons. First is their inherent lack of water solubility. Second is the large steric effect due to bulky substituents. Experimentally, the low solubility can be circumvented by using a mixed solvent system, one component of which is organic but water-soluble. 2-Propanol is convenient in this respect and also does not react with base.

Although a direct comparison with water as solvent cannot be made when using water/2-propanol systems, one usually finds that as the solvent polarity increases, the rate constant of the hydrolysis reaction generally increases. Thus, if the reaction is carried out in a series of solvent mixtures with varying water composition, the results may be extrapolated to that for pure water.

Safety Issues—Wear safety glasses at all times in the chemistry laboratory.

Procedure

1. A 70% 2-propanol/water (v:v) solution will be provided. The plasticizer studied in this experiment, dimethylphthalate (DMP) is very soluble in 2-propanol.

2. Make up 50 mL each of 0.40 M DMP and KOH in 70% 2-propanol. Mix the two solutions together in the reaction vessel, a 200 mL tall form beaker. Measure the conductance as quickly as feasible. This first value is L_0. Measure the temperature of the system. A simple apparatus for kinetics measurements is shown in Figure 19-1.

3. Measure the conductance of the reaction system at 3–5 minute intervals for 90 minutes. The temperature should be measured after each conductance measurement. A conductance meter with a temperature probe is very convenient for these measurements

Waste Minimization and Disposal

1. The waste is about 100 mL of 2-propanol, water, dimethylphthalate, and potassium hydroxide. The hydroxide can be neutralized with 6 M HCl (requires about 2 mL). The reaction mixture should be transferred to a waste container at the completion of the experiment. This will be disposed of by using a commercial facility.

2. When the experiment is complete, the glassware is rinsed with 2-propanol and the rinsings added to the waste container.

Data Analysis

1. Prepare a table that contains the raw data, including time, conductance, and temperature, which should remain nearly constant.

2. Use the experimental data to plot $(L_0 - L_t) / L_t$ (ordinate) versus t. If possible use a computer to generate the plot and to carry out a least-squares analysis from which the slope and correlation coefficient are obtained.

3. Use the slope from the previous plot to calculate the rate constant for the reaction. Compare the rate constant with values for fairly fast and fairly slow second-order reactions (refer to general chemistry texts or books on chemical kinetics, such as Ref. 1).

Figure 19-1 Apparatus for Kinetics Measurements Using Conductance

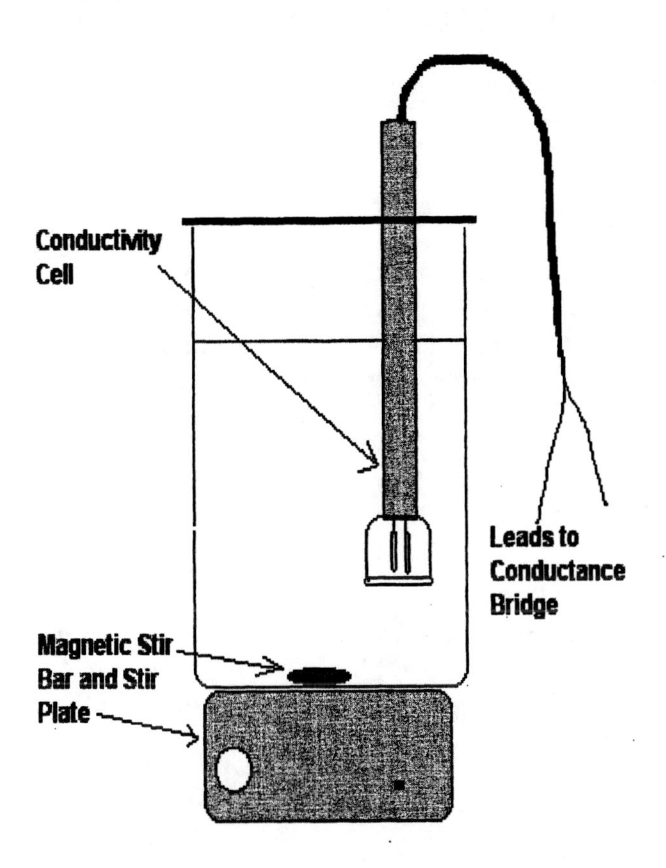

Supplemental Activity

1. If several groups are doing the experiment concurrently, different solvent compositions may be used. This will allow the rate constants to be compared by plotting the rate constant (ordinate) versus the composition of the solvent system.

2. If several groups do the experiment at the same time, different temperatures can be used. This will allow the activation energy for the reaction to be obtained.

3. Many other esters of environmental significance are readily available, and the effect of structure on the rate constant can be studied. This may be more desirable for a student with more chemistry background—especially in physical or organic chemistry.

Questions and Further Thoughts

1. How will the reaction rate change if the methyl group in dimethylphthalate is replaced by a butyl group, considering the steric effect? Computer software showing reactants as space-filling three-dimensional structures will give you a good picture of the size effect on the rate of reactions.

2. How would chlorination of the methyl group affect the rate constant? See the previous comments.

Notes

1. The reaction given is slow at 25°C, and carrying it out at higher temperatures may be more interesting.

2. The initial points in the plot done in the Data Analysis section may be off the straight line generated. After getting the initial conductance you may want to wait several minutes before taking readings, or you may wish just to include them and get a correlation coefficient, and then carry out a least-squares analysis without those points.

Literature Cited

1. A. A. Frost and R. G. Pearson, *Kinetics and Mechanism*, Wiley, New York, 1953.

2. M. Grayson, Executive Ed., *Kirk-Othmer Encyclopedia of Chemical Technology,* Wiley Interscience, New York, 1982, Vol 18, pp. 111-183.

3. J. M. Wilson, R. J. Newcombe, A. R. Denaro, and R. M. W. Rickett, *Experiments in Physical Chemistry,* 2nd ed., Pergamon Press, London, 1962.

Experiment 20

Properties of Detergents: Surface Tension Measurement of Critical Micelle Concentration

Objectives—The objectives of this experiment are:

1. To introduce you to the fundamental properties of surfactants
2. To illustrate the properties of surfactants by measuring surface tension
3. To gain familiarity with the critical micelle concentration
4. To gain familiarity with the surface tensions of natural waters

Introduction

Soaps, detergents, and surfactants Ordinary soap is a mixture of the sodium salts of monoprotic carboxylic acids containing 10–18 carbon atoms. Shorter chain salts are unable to emulsify oil in water, whereas longer chain salts are too insoluble to function as effective detergents. Two important components of ordinary soap are sodium palmitate, $C_{15}H_{31}COONa$, and sodium stearate, $C_{17}H_{33}COONa$.

Soaps are made by reacting a strong base with fats, which are glycerides (esters of glycerol). The decomposition of an ester to produce a carboxylic acid and an alcohol is called saponification (the opposite of esterification), the original meaning of the word is soap-making (Latin: *saponis*). These reactions were discussed in Experiment 19.

A major problem with soaps is that they precipitate in hard water; that is, water that contains calcium, magnesium and/or iron ions. Some useful properties of soaps are:

1. They are excellent for lathering and cleansing.
2. They are efficient for laundry and industrial applications.
3. They are biodegradable in wastewater treatment plants and in natural waters.

Soap is one type of surfactant, an organic compound that has within the same molecule two dissimilar structural groups, one a water-soluble (hydrophilic) moiety and the other a water-insoluble moiety (hydrophobic group). A surfactant affects the properties of a solvent to a much greater extent than expected on the basis of its very small concentration. This marked effect is due to (1) adsorption of the surfactant at the solution interfaces and (2) micelle formation in the bulk of the solution. A surfactant increases the wetting ability of water, allowing it to loosen, solubilize, and/or emulsify soils, and hold the loosened dirt in suspension, preventing it from settling back into the cleaned items.

There are four groups of surfactants, based on the nature of their hydrophilic group: anionic, cationic, nonionic, and amphoteric (a combination of anionic and cationic). Generally, anionic and nonionic surfactants are relatively nontoxic to mammals. They are in the same category as sodium chloride and sodium carbonate. Cationic surfactants are only slightly more toxic.

The molecular weights of surfactants start at about 200 (for sodium hexylsulfate) and extend into the thousands. Many surfactants are polydisperse; that is, they vary in chain length or in some other structural aspect.

A surfactant usually displays its greatest surface activity when it is present close to its limit of solubility, which is between 0.1 and 10 mM for efficient surfactants.

A **detergent** is a cleansing agent and soap is one type of detergent. Detergents lower the surface tension of a solvent to about one-third of its original value. Detergents are usually mixtures, and consist of a surfactant, a chelating agent or builder (to complex ions that cause hardness), a bleach, an optical brightener, a perfume, and inert fillers, such as sodium sulfate.

Syndets are synthetic detergents and have been available since 1930.

Wetting Agents Substances that reduce surface tension are called wetting agents. Wetting power is the capacity of a liquid to wet a solid, a process that displaces dirt or oil from the solid. Wetting agents are used to (1) wet pesticide powders to form aqueous suspensions that can be sprayed, (2) convert oil–wet sand to water–wet sand to release the oil, allowing it to migrate to a producing well, and (3) penetrate corroded or oily metal surfaces by cleaning solutions.

Light-duty (or unbuilt) detergents, were developed for hand-washing dishes and light laundry. These liquids would suds too much in washing machines and thus are not used for the heavy family wash.

All purpose (or built) detergents are used for washing clothes. They contain a "builder," which formerly was usually a complex phosphate. Builders enhance the cleaning power of a detergent, making it capable of removing soil from fabrics.

Phosphates The builder, formerly complex phosphates, has several important functions in detergents.

1. It increases the efficiency of the surface-active agent by complexing ions that cause hardness.

2. It keeps dirt particles in suspension.

3. It increases alkalinity, which is needed for cleaning action.

4. It emulsifies oily and greasy solids.

5. It acts as a bactericide.

Phosphates in wastewater:

1. Break down into *ortho*-phosphate by hydrolysis in sewage treatment plants and receiving waters

2. Do not interfere with waste treatment processes

3. Already exist as a part of the natural environment

In the past, most detergents contained 35–50% sodium tripolyphosphate, $Na_5P_3O_{10}$.. Once metals become complexed with polyphosphates, they remain stable and dissolved in water. Excess polyphosphate eventually hydrolyzes to *ortho*-phosphate, which is dispersed into environmental waters. Phosphorus is essential for the growth of algae and aquatic plants. However, when phosphorus levels become too elevated, plant life in the water, mainly algae, gets out of control and makes navigation impossible. The enrichment of natural waters with nutrients is called eutrophication.

Phosphorus is not a pollutant in the usual sense since it is involved in natural processes needed to sustain the life of freshwater systems. It enables plants to grow in the water, providing food for fish. Phosphates in the environment also originate from the weathering of phosphorus-containing minerals, human and animal wastes, industrial wastewaters, and runoff from agricultural land, due to animal wastes and fertilizer usage..

Many detergents are now free of all phosphates, and the builders used include washing soda (sodium carbonate), sodium silicate (waterglass), and sodium citrate. Detergents that still use phosphates use from 0.5 to about 2% polyphosphate.

The Biodegradability of Detergents A biodegradable material can be broken down and used as food by the bacteria commonly found in sewage, surface waters, and soils. All household detergents in the United States have been biodegradable since 1965. Foaming incidents since then were caused by raw sewage, industrial wastes, waxes from foliage, etc. The loss of surfactant characteristics by biodegradation can be measured using properties such as surface tension or foaming tendency.

Synthetic surfactants were developed to get around the problem of soaps precipitating in hard water. The first of the synthetic detergents were the branched alkyl benzene sulfonates (ABSs). The surface activity of the sulfonate group, SO_3^-, does not depend on pH (which can precipitate out soaps at low pH) or the presence of heavy metal ions. Although the ABSs replaced soaps for many uses, they were found not to be biodegradable.

Linear hydrocarbon chains biodegrade faster than branched chains. The degrading of surfactants is primarily accomplished by bacteria that oxidize linear hydrocarbon chains two carbons at a time, and a branch in the chain interrupts the degradation.

The linear sulfonates, (LASs), replaced the ABSs in 1965. An important LAS is sodium lauryl sulfate, and coconut oil, a glyceride, is used to make this surfactant. The fat is first reduced to a mixture of alcohols through the reaction:

$$\begin{array}{lcl} RCOOCH_2 & & RCH_2OH \\ | & \text{Reduction} & + \\ R'COOCH & \longrightarrow & R'CH_2OH \quad + \quad CH_2-CH-CH_2 \ (OH\ OH\ OH) \\ | & & + \\ R''COOCH_2 & & R''CH_2OH \end{array}$$

(a glyceride) (mixture of alcohols) (glycerine)

This reaction yields a high percentage of lauryl alcohol, n-$C_{12}H_{25}OH$, which is then converted to a sulfate ester with sulfuric acid, followed by treatment with base to form sodium lauryl sulfate (SLS).

$$\underset{\text{Lauryl alcohol}}{CH_3(CH_2)_{10}CH_2OH} + HOSO_2OH \longrightarrow CH_3(CH_2)_{10}CH_2OSO_2OH + H_2O$$

Theory—Surfactants form aggregates of molecules or ions called micelles when the concentration of the solute exceeds a limiting value called the **critical micelle concentration** (CMC). The CMC is the concentration at which micelles begin to form and its value varies with the solute-solvent system. A micelle is a submicroscopic fundamental component of a surfactant. Since a surfactant has two different types of groups, hydrophobic and hydrophilic, the hydrophobic groups tend to associate with each other when they are in the body of the solution. The hydrophilic groups, on the other hand, associate with the aqueous environment. Both properties can be realized by forming a micelle, as shown in Figure 20-1.

Figure 20-1 Representation of a Micelle

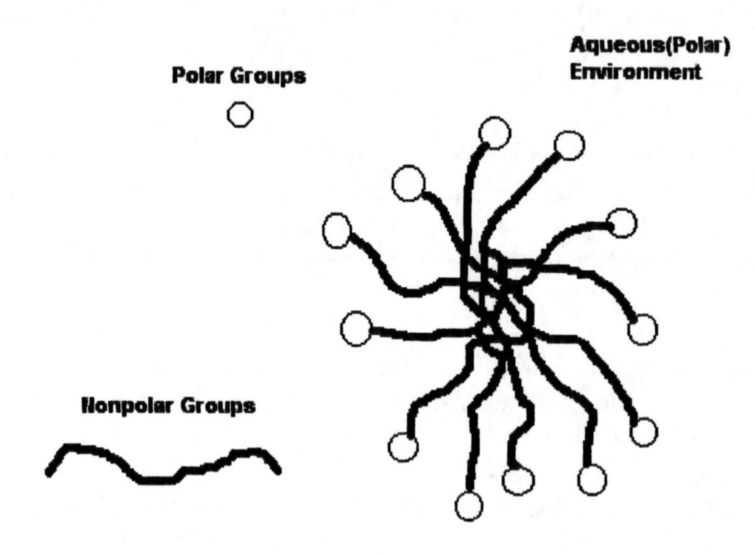

The spontaneous dissolving of a normally insoluble substance by a relatively dilute solution of a surfactant is called micellar solubilization. Solubilization and detergency begin at the CMC and increase with increasing surfactant concentration.

The CMCs of nonionic surfactants are generally much lower than those of ionic surfactants of comparable hydrocarbon chain length. The CMCs of anionic surfactants in aqueous solutions are about 10 mM; for nonionic surfactants they are about 0.1 mM. For a series of surfactants, where the hydrophilic group remains the same, CMC values decrease with increasing chain length of the hydrophobic group for both ionic and nonionic surfactants. The CMCs of anionic surfactants increase with increasing temperature, whereas the CMCs of nonionic surfactants decrease with increasing temperature.

The most frequently determined properties of a surfactant are its surface tension and its CMC. If the properties of a surfactant solution are plotted as a function of the concentration of the surfactant, the properties will usually vary linearly with increasing concentration up to the CMC, at which point there will be a break in the curve. The CMCs of surfactants have been determined by the:

1. Break in the surface tension versus log (concentration) curve

2. Break in the turbidity versus concentration curve

3. Iodine-solubilization method

4. Conductance versus concentration curve (for ionic surfactants)

5. Viscosity versus concentration curve

Surface Tension and Its Measurement One of the most characteristic features of a liquid is its surface tension. It is due to unbalanced intermolecular forces at the surface of the liquid. Most substances, when dissolved in water, reduce the surface tension of water. Since most substances have a surface tension less than water, when they are uniformly distributed at the water surface, they result in a smaller surface tension. Detergents decrease the surface tension by a much greater amount than expected on the basis of the small amounts present. A few substances, mainly ionic salts, increase the surface tension of water, but only to a small extent.

The surface tension, γ, may be defined as the amount of work necessary to extend the surface of a liquid by unit amount,

$$\gamma = \text{work} / \Delta A \tag{20-1}$$

where A is area. The units of surface tension are erg/cm^2, or dyn/cm (joules per square meter, J/m^2, or N/m in the S.I. system). The surface tension of water is 72.75 dyn/cm at 20°C and that of diethyl ether is 17.01 dyn/cm at 20°C. The large surface tension of water is due to the very strong hydrogen-bonding forces present, whereas the forces in ether are chiefly dispersion.

Several methods are available for measuring the surface tension. In the capillary-rise method, the height to which a liquid rises in a glass capillary tube is measured. In the weight-drop method, the weight of a drop that falls from a tube of known radius is determined. The ring method is based on measuring the force necessary to pull a metal ring from the surface of a liquid. In this experiment we shall use the capillary-rise method.

The Capillary-Rise Method A liquid tends to assume a shape of minimum area. If a vertical capillary tube is dipped into a liquid that "wets" the tube, a film of the liquid will run up the capillary wall. Then, to reduce the surface area, the liquid rises in the tube until the force of gravity on the liquid column balances the tension at the circumference. The simplest equation for calculating surface tension by capillary rise is,

$$\gamma = rh\rho g / 2 \qquad \textbf{(20-2)}$$

where r is the capillary radius, h is the rise in the capillary, ρ is the density of the liquid, and g is the acceleration due to gravity. More accurate equations can be used that take contact angle into account, but Eq. 20-2 is satisfactory for this experiment. The capillary rise method is probably the most accurate way to measure surface tension.

Safety Precautions

1. Safety glasses must be worn at all times in the chemistry laboratory.

2. Wear neoprene gloves when handling concentrated nitric acid. Use nitric acid only in a fume hood. This acid is highly oxidizing. Also avoid contact of nitric acid with organic matter since it may be explosive.

Procedure

A: Calibration of the Capillary

Capillary tubes are cleaned, if necessary, by soaking in concentrated nitric acid. They should be stored in pure water. Carry out the measurements as follows, referring to Figure 20-2.

1. Place enough sample in the large outer tube so that when the capillary-rise scale is placed inside, the outer liquid level lies between 0 and 1 cm. The capillary can also be adjusted to some extent. To avoid frothing, which will make it difficult to read the level of the outer liquid meniscus, gently pour the liquid down the inside wall of the outer tube.

2. Rinse the capillary with sample by forcing solution into the capillary using a bulb on the side arm.

3. Adjust the height of the liquid above its equilibrium value using the bulb and measure the final height. Repeat both steps. A good light and a magnifier may be needed to see the liquid level within the capillary. (In each measurement, determine the distance between the lower meniscus in the outer tube and the upper meniscus in the capillary.)

4. Average the values of two readings to obtain a value of h. A difference of more than ±0.2 mm in the readings indicates a dirty capillary.

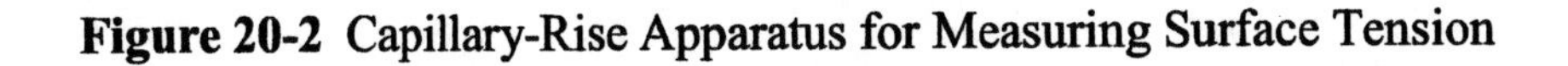
Figure 20-2 Capillary-Rise Apparatus for Measuring Surface Tension

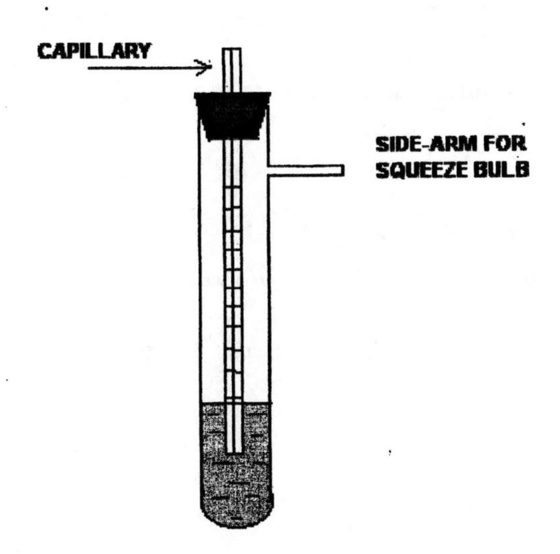

Although the capillary usually comes with a known diameter (about 0.5 mm), for precision work the radius should be accurately determined, using water to calibrate. The values for the surface tension and density of water are given in Table 20-1. From the measured height and the temperature of the water, Eq. 20-2 is used to calculate the value of r. Two measurements of h should be made and two values of r calculated and the average value of r used in subsequent calculations.

When other liquids are to be examined in a tube that has been used, rinse the capillary with the new liquid outside the apparatus and then reassemble the apparatus. Alternatively, rinse the capillary with a volatile solvent and then dry the capillary with a stream of gas.

Table 20-1 The Surface Tension and Density of Water at Various Temperatures

Temperature (°C)	Density (g/mL)	Surface Tension (dyn/cm)
20	0.99823	72.75
25	0.99707	71.97
30	0.99567	71.18

Source: R. C. Weast, Editor-in-Chief, *CRC Handbook of Chemistry and Physics*, 59th ed., CRC Press, Inc., Boca Raton, FL, 1978.

B: Determining the CMC of an Ionic Surfactant

1. Prepare a stock solution of the detergent in DI water. Its concentration depends on the CMC of the surfactant under study. For a detergent like sodium lauryl sulfate (SLS), with a CMC in the vicinity of 0.01 M, a 0.2 M stock solution is convenient. This is near the solubility limit, and sudsing begins to become a problem at such high concentrations.

2. Use pipets and volumetric flasks to prepare 100 mL of each of 10 solutions with varying concentrations of surfactant by appropriate dilutions of the stock solution. About five solutions should have a concentration less than the CMC, and the other five should have concentrations greater than the CMC.

3. Measure the capillary rise of each solution using the procedure in Part A.

C: Measuring the Surface Tension of Environmental Samples

1. Measure the capillary rise of four environmental samples—a stream, a pond, a well, and seawater (if possible). These samples **must be filtered** to avoid clogging the capillary.

2. If time permits, the density of each sample is determined. For dilute solutions of surfactant, the density will be about the same as that for pure water.

Waste Minimization and Disposal

1. If concentrated nitric acid is used to clean capillaries, it can be stored in a bottle for further use and need not be discarded. If it is eventually discarded, it is neutralized with dilute base to a pH between 3 and 11, after which it can be flushed down the laboratory drain (after approval of the instructor).

2. The surfactants chosen for study should be biodegradable and thus can be flushed down the laboratory drain.

3. The environmental water samples studied in this experiment can be samples collected for previous experiments, or if collected especially for this experiment, they can be used for subsequent experiments. Very little sample is needed for a surface tension measurement by capillary rise. A 50 mL sample is sufficient for rinsing and triplicate measurements. If the samples are to be discarded, they can be flushed down the laboratory drain if they are not hazardous wastes. Otherwise, they can be collected and disposed of according to the practice suggested by your school's safety officer.

Data Analysis

1. Use a table to report your calibration data, the capillary rise, temperature, density and the radius of the capillary. Compare your calculated capillary radius with a given value, if available, or with previously measured values.

2. Calculate the surface tension for each solution studied. If the densities of the solutions were experimentally measured, list their values in a table.

3. Prepare a plot of surface tension (ordinate) versus log (surfactant concentration). Draw the best possible straight lines through the points on each side of the CMC and find their intersection, which corresponds to the CMC. Report this value. Also try a plot of surface tension versus the concentration and discuss the result.

4. Compare your experimental CMC with values expected for ionic surfactants. Discuss the differences and experimental errors.

5. Calculate and discuss the surface tension results for the environmental samples. From the values obtained, what can you conclude about the nature of the dissolved solids in the samples?

Supplemental Activity

1. The CMC can be determined by measuring the viscosity of the surfactant solutions and plotting the viscosity (as ordinate) versus either log (concentration) or simply the concentration. An Ostwald-Cannon-Fenske viscometer is used and the size chosen so that it takes at least several minutes for the liquid to flow through the capillary. The viscometer is calibrated with water and the following equation is used to calculate the viscosity of the surfactant solutions, $\eta_s / \eta_w = \rho_s t_s / \rho_w t_w$, where the ρ's are densities and the t's are the efflux times in the viscometer. If the densities of water and surfactant solutions are not much different, the densities cancel, leaving a very simple relationship.

2. A very quick method for measuring the CMC of an ionic surfactant is to measure the conductance of the series of surfactant solutions. Plotting the conductance (as ordinate) versus concentration locates the CMC.

3. If the densities of the solutions are accurately measured, it would be of interest to plot these versus concentration to see if the density could also be used to locate the CMC. Generally densities are insensitive to small amounts of impurities (or solutes).

Questions and Further Thoughts—Other physical properties may also be used to determine the CMC of a surfactant. One might try, for example, the refractive index, which is quite easy to measure.

Notes

1. Frothing can be very troublesome in this experiment. When making the solutions they should not be overly shaken and should be given some time to settle.

2. A thermometer magnifier or a cathetometer telescope can facilitate the measurements. A magnifier "borrowed" from a melting-point apparatus aids significantly in the measurements of the liquid level in the capillary.

Literature Cited

1. C. Baird, *Environmental Chemistry*, W. H. Freeman, New York, 1995.

2. D. Brown, Detergents and the environment, *Chem. Ind.(London), 2,Feb.1974(3):93-95.*

3. M. Grayson, Executive Ed., *Kirk-Othmer Encyclopedia of Chemical Technology*, Wiley-Interscience, New York, 1982, Vol.21, pp. 162-181, Vol. 22, pp. 332-432.

4. David R. Lide, Editor-in-Chief, *CRC Handbook of Chemistry and Physics*, 77th ed., CRC Press, Inc., Boca Raton, FL, 1996.

5. S. E. Manahan, *Environmental Chemistry,* 5th ed., Lewis Publishers, Chelsea, MI, 1991.

6. M. J. Schick, Ed., *Nonionic Surfactants,* Marcel Dekker, New York, 1967.

7. K. Shinoda, T. Nakagawa, B. I. Tamamushi and T. Isemura, *Colloidal Surfactants,* Academic Press, New York, 1963.

8. D. Stigter, Micelle formation by ionic surfactants. II. Specificity of head groups, micelle structure, *J. Phys. Chem. 78*(24):2480–2485 (1974).

Air Sampling and Analysis 4

Experiment 21

Introduction to Air Sampling: Particulates in Urban Air

Objective—The objective of this experiment is to introduce you to the basic techniques used to collect substances present in the atmosphere, with an application to the determination of particulates.

Introduction—Air studies are seldom performed in basic environmental science laboratory experiments due to the intrinsic difficulty of collecting representative samples and also due to the small concentrations of pollutants present. The low concentrations of air contaminants necessitates taking a large sample or passing air through a collection device for a longer period of time.

At a minimum, an aspirator can be used to draw air through a collection device, such as a Büchner funnel with filter paper. Other methods involve the use of "bubblers," which contain colorimetric or precipitating agents.

Air Pollution.

A point source of pollution (air or water) is one that has a fixed geographic location. These are generally the easiest to identify (you can point to them) and to sample. Large point sources of pollution (such as municipal sewage treatment plants or paper companies) are required to obtain permits to pollute. Several types of permits exist, depending on the amount of the pollutant, whether the facility has a specific monitoring requirement (e.g., air toxins), etc. There are several hundred point sources of air pollution in most large urban areas, including the common (residential fireplaces, mom-and-pop bakeries, dry cleaners, flame-broiled and open flame BBQ and steak houses, incinerators—including mortuaries and hospitals) and the temporary (backyard cook-outs, accidental fires, etc.).

A nonpoint source of pollution (air or water) is one that is geographically diffuse; it has no definite point of origin. Examples of nonpoint sources of air pollutants are sea salt aerosol, products of reactions of gases in the air to form secondary pollutants (photochemical smog), and vehicle emissions.

Whether a city is likely to have an air quality problem is determined by three factors: local sources of emissions (and the type of pollutant emitted), topography (the terrain of the land), and meteorology. There is little that can be done about the topography or meteorology of a locale except to understand it. The meteorology of an area is determined by its location—specifically, its proximity to the ocean and its latitude. A few points on the effects of meteorology on air quality bear mentioning. A plot of the direction the wind blows over time is called a wind rose, with the predominant wind direction showing up as a point on a circular graph. In some areas of the United States the wind rose shows that the wind blows about equally from all directions on a yearly average. In the summer months, warm temperatures and stable high-pressure zones in the Atlantic Ocean contribute to air stagnation. With no wind blowing, pollutants build up in the atmosphere, where they can "cook" in the higher ultraviolet radiation, especially in areas close to the equator.

An atmospheric inversion occurs when colder air is trapped below a layer of warmer air. The result is a "lid" beyond which air emissions are not transported. Some areas of the United States experience inversion conditions nearly 50% of the time. Inversions result from the difference in warming and cooling rates of air and land.

Most point sources do not change their emissions at any time in the day; however, the intensity of the odor is stronger at different times of the day, particularly in the early morning hours. The explanation is as follows: water-soluble gases, such as the reduced sulfur species from paper mills, dissolve in the water (dew) that condenses at night onto grass, leaves, and the ground. As the early morning sun heats this layer, the gases vaporize (gases are less soluble at higher temperatures) and add to the emissions of the morning. This results in an increased atmospheric concentration of "smelly stuff."

Lung Cancer. The single largest factor determining whether a person will develop lung cancer is life span. The time it takes for a person to develop lung cancer varies with the individual; however, the lag time between exposure to a causative agent and development of the disease can be 20–40 years. Thus, it is not often easy to tell what caused the disease (or where or when). Further complicating the issue is the wide variability in the susceptibility of individuals to cancer in general.

One factor that seems to play a role in the development of cancer is heredity. Some people appear to be "immune" or resistant to cancers. Race and sex differences show up in the frequency of many types of cancers, including lung cancer. Among those agents definitely linked to lung cancer are smoking habits.

Indoor Air Pollution

Humans have been polluting indoor air since cave people started the first fire inside. Although indoor air quality is a relatively recent field of research, early caveman contended with excess particulate matter on occasion. In modern times, 70% of the U.S. population lives in metropolitan areas (where, presumably, the air is worse) and spends 90% of their time indoors. The Environmental Protection Agency and other federal, state and local agencies have begun to address the problems of "bad air" in buildings—the sick building syndrome. As a result of energy efficiency efforts in the 1970s, architects and builders began to seal "leaks" of air into and out of buildings. Windows were sealed and circulated air replaced older, once-through systems. Consequently, the air was trapped inside to collect whatever vapors, particles, organisms, and gases that were in the building.

The following lists types of pollutants found in the home and their location:

Kitchen: cooking odors and greases; cleaners and other chemicals under the sink; asbestos from vinyl tiles or pipe insulation

Furnace Room (basement, closet, etc.): carbon monoxide from poorly maintained furnace, vapors and gases from volatile materials (paints, solvents, etc.); particulate matter not captured due to dirty furnace filter

Bathroom: chlorinated hydrocarbons from chlorinated water (released during shower); mildew and mold; volatile chemicals from deodorants, disinfectants, etc.; ingredients and propellants from personal hygiene products (used in the smallest room of the house)

Closets: *para*-Dichlorobenzene or naphthalene (mothballs); tetrachloroethylene (and other dry cleaning solvents); mold, mildew, and fungi in shoes

Garage/Workshop: gasoline; solvents from paint, thinners, strippers, etc.; particulate matter; carbon monoxide and other fumes from the automobile, generated during engine start-up

Living/Dining Room: volatile compounds from stain removers; organics from synthetic fibers (carpets, upholstery, etc.); formaldehyde from particleboard and blown-foam insulation; insecticides; carpet and air fresheners

Priority Pollutants Large cities are required to monitor six chemical species of nonpoint pollutants in order to satisfy the Environmental Protection Agency that they are complying with state and federal regulations. The Environmental Protection Board of Jacksonville (Florida) Annual Report (similar agencies exist in other cities) lists the pollutants monitored, the standards to be met, the maximum concentrations found at various sampling sites, and a nonspecific scale called the Air-Quality Index, which indicates how the air quality is overall (good, moderate, or unhealthy). The six priority pollutants monitored and the standards are given in Table 21-1.

Table 21-1 Standards for Priority Pollutants

Pollutant	Averaging Time	Federal Standard	Florida Standard
Carbon monoxide	8 hr	9 ppm	9 ppm
	1 hr	35 ppm	35 ppm
Lead	3 mo	1.5 $\mu g/m^3$	1.5 $\mu g/m^3$
Nitrogen dioxide	Annual	50 ppb	50 ppb
Ozone	1 hr	120 ppb	120 ppb
	3 yr	< 1 ppb	< 1 ppb
Particulate matter (PM-10)	Annual	50 $\mu g/m^3$	50 $\mu g/m^3$
Sulfur dioxide	Annual	30 ppb	20 ppb
	24 hr	140 ppb	100 ppb
	3 hr		500 ppb

Particulates in the Atmosphere

Particulates are very small solid or liquid particles suspended in air and are not usually visible to the naked eye. Atmospheric particulates extend in size from about 0.5 mm in diameter (the size of sand) down to molecular dimensions and may be composed of a wide variety of substances. Although the term particulates has come to stand for particles in the atmosphere, particulate matter (PM) is the preferred term.

There are a number of common names for atmospheric particles: "dust" and "soot" refer to solids, "mist" and "fog" refer to liquids, fog denoting a high concentration of water droplets.

Particulate matter constitutes the most obvious form of air pollution. Particles are usually suspended in the air near pollution sources such as industrial and power plants and highways.

Particles with diameters less than 2.5 μm are called fine particulates and usually remain airborne for days to weeks. Particles with diameters greater than 2.5 μm are called coarse particulates and settle out rapidly. In addition to sedimentation, particles can be removed from the atmosphere by absorption into falling raindrops. The concentration of particulates in air is reported in units of $\mu g/m^3$. A common air quality standard for total suspended particulates (TSP) is 75 $\mu g/m^3$. However, since only fine particles are of most concern to human health, a more appropriate index is the concentration of particulates with less than some particular diameter. The concentration in air of particulate matter with diameters less than 10 μm is described by the notation PM_{10} . The PM_{10} is about one-half the TSP in most cities. The air standard for PM_{10} in the United States is 150 $\mu g/m^3$. A PM_{10} greater than this value, averaged over a 24-hour period, appears to adversely affect human health. Epidemiological studies associate particulates more directly with poor health than any gaseous pollutant.

Inhaled fine particles usually travel to the lungs and become adsorbed onto cell surfaces. Such particles are said to be respirable. When inhaled, coarse particles are efficiently filtered by the nose and throat and generally do not get to the lungs. The respiratory system possesses mechanisms for the elimination of inhaled particles. Particles cleared from the respiratory tract are to a large extent swallowed into the gastrointestinal tract.

Atmospheric particulates are of concern because they enter the lungs, blocking and irritating air passages, and in themselves exert toxic effects. Emphysema is associated with the accumulation of particles in the lungs.

Industrial processes contribute the most particulate emissions to the atmosphere (about 37%). Additional inputs come from solid waste treatment, stationary source fuel combustion, and transportation. Natural sources are also important, and wind-blown topsoil is an important example.

Large solid particles, such as those that make up "dust," originate mainly from nonchemical processes. Dust is produced by natural sources, such as volcanic eruptions, and human activity, including the crushing of rocks in quarries and land cultivation.

Minerals contribute a portion of the particulate matter in the atmosphere. Since many of the large particles in atmospheric dust originate as soil or rock, it is expected that their elemental composition will be similar to the composition of the earth's crust. Therefore, they contain large amounts of aluminum, calcium, silicon, and oxygen, often in the form of silicates.

Among the constituents of inorganic particulates are salts, oxides, nitrogen and sulfur compounds, and metals. Trace metals that typically occur at levels greater than 1 $\mu g/m^3$ in particulates are aluminum, calcium, carbon, iron, potassium, sodium, and strontium. In areas near and over the oceans the concentration of solid NaCl in medium-sized particles is high.

When high-ash fossil fuels are burned, small particles of fly ash enter furnace flues and are collected in a stack system. However, the very small particles of fly ash are released to the atmosphere where they can damage human health and plants and also reduce visibility. The most important components of fly ash are the oxides of aluminum, calcium, iron, and silicon. Other elements present include magnesium, sulfur, titanium, phosphorus, potassium, and sodium. Elemental carbon (soot, carbon black) is also a significant component of fly ash.

Soot particles are a problem since they can adsorb large amounts of toxic substances on their irregular surfaces. Such particles originate in diesel exhausts, wood-fire smoke, furnaces, incinerators, power plants, and industrial processes resulting in incomplete combustion. Elevated levels of PAHs of up to about 20 $\mu g/m^3$ are found in the atmosphere, chiefly adsorbed to soot.

Heavy metals are usually transported in the atmosphere as species adsorbed on or absorbed in suspended particulates. More than one half of the heavy metal input into the Great Lakes in the past was due to deposition from the atmosphere.

Theory

The Composition of Organic Particulates. Organic atmospheric particulates contain a wide variety of compounds. The analysis of these particles is done by first collecting them on a filter, followed by extracting with organic solvents, and then fractionating into neutral, acid, and basic groups, and finally analyzing for specific compounds.

1. Neutral Group: Composed chiefly of hydrocarbons, including aliphatics and aromatics and oxygenated fractions. The aliphatic fraction contains a large percentage of long-chain hydrocarbons, predominantly having 16–28 carbon atoms. The aromatic fraction contains PAHs that are often carcinogenic. Aldehydes, ketones, epoxides, peroxides, esters, quinones, and lactones are among the oxygenated neutrals, some of which can be mutagenic or carcinogenic.

2. Acidic Group: Contains long-chain fatty acids and nonvolatile phenols.

3. Basic Group: Contains *N*-heterocyclic hydrocarbons.

Air Sampling. Automobile air filters, furnace/air conditioner filters, and vacuum cleaner bags collect particulate matter by simple filtration. One of the earliest air pollution studies conducted in New York City had health workers offering citizens free air conditioner filters in exchange for used ones, which were shaken and the contents observed

Quantitative analysis of the particulate matter in the atmosphere is accomplished by drawing air through a filter for a known period of time at a known flow rate. The mass of the filter is determined before and after sampling, and the concentration of particulate matter is reported in terms of micrograms per cubic meter ($\mu g/m^3$) of air sampled.

If an oil-free vacuum pump is available, samples can be obtained by drawing air at about 17 L/min through a 37 mm diameter filter for 24 hours. The equipment needed is listed in Table 21-2.

Table 21-2 Particulate Sampler Parts and Functions

Equipment	Explanation
Oil-free pump	Any small pump which draws in enough air
Filter holder	Gelman Air Monitoring Cassette, 3-Piece, reusable
	Replace membrane filter with paper or glass fiber
Filter	Quartz Fiber (or Whatman 40)
Tubing	Connects filter to pump
Flowmeter	For example, Dwyer rotameter

Safety Issues—Safety glasses must be worn at all times in the chemistry laboratory.

Procedure

Collection and Comparison of Indoor and Outdoor Particulates In this experiment 48-hour air samples will be collected inside the laboratory and on a roof outside the laboratory to compare total particulates. The exact location of the samplers must be noted. A rain shield is installed over the outside sampler. After collection, the filters should be dark and the relative darkness of the two filters is noted.

A: Preparation of the Quartz Fiber Filters

1. Place 37-mm-diameter quartz-fiber filters in a 150 mL beaker, add 25 mL of high-quality cyclohexane, and cover with aluminum foil. Place the beaker in an ultrasonic bath and sonicate for 15 minutes on high.

2. Remove the filters with forceps and put them onto cyclohexane-rinsed watch glasses for 5 minutes to air dry. Then dry for 10 minutes in a drying oven at 110°C. Allow to cool in a desiccator and then weigh to the nearest 0.1 mg. (If available, a microbalance should be used to determine the mass of the filter before and after sample collection.)

B: Assembly of the Apparatus

1. Assemble the apparatus by arranging the following into a chain: air-sampling pump, tube, filter holder, and timer. The filter holder is contained in a rain shield, which can be made from a one-gallon water bottle with the bottom cut off. This prevents rain from coming into contact with the filter and should be used on the outside collector. (See Figure 21-1.)

2. Prior to sample collection, the air flow rate is measured and recorded. After sampling, a final flow measurement must also be made. The average flow rate is the sum of the two measurements divided by two. It is always an assumption that the material of interest is deposited uniformly across the sampling period. From the measured weight gain and volume of air, the mass of particulates per unit volume is calculated.

Figure 21-1 Air-Sampling System for Particulates

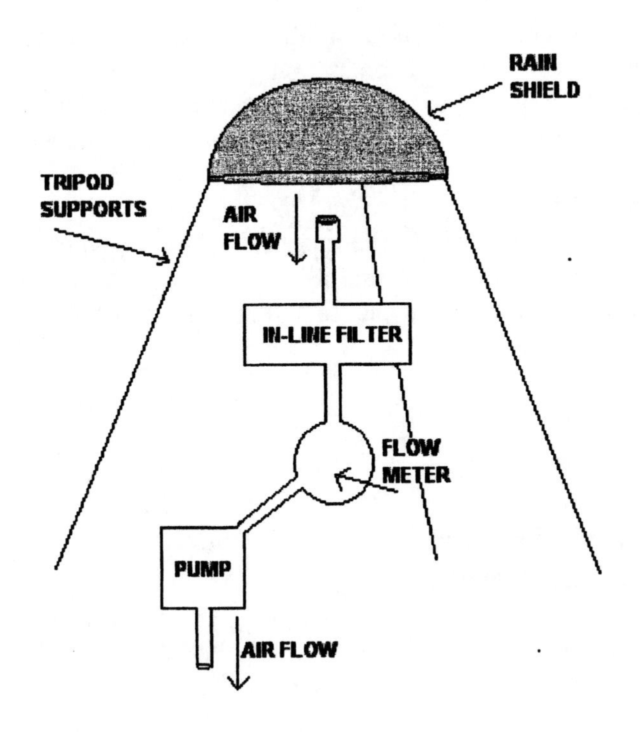

3. Attach the sample holder containing the preweighed filter to the tube from the intake of the pump.

4. Attach the flow meter, which is connected to the holder cap by means of a short tube, to the filter holder. NOTE: Work as quickly as possible for the next step.

5. Turn on the pump. Holding the flow meter vertically, read the center of the ball as the Initial Flow Rate. Turn the pump off.

6. The flow rate should be between 15 and 20 L/min. If it is not, consult the instructor who will adjust the pump.

7. Position the rain shield outside the window of the laboratory.

8. When ready to begin sampling, note the Start Time, start the pump, and record any important meteorological conditions (especially rain).

9. After approximately 48 hours, turn off the pump (noting the final time) and bring the rain shield containing the sample inside. Repeat Step 3, and record the Final Flow Rate.

10. Note the color and extent of darkness of the filter. Put the filters into a desiccator for 0.5 hour and then weigh the filter. Store it in a dessicator until it is used in any subsequent experiments. Using the desiccator reduces the amount of sorbed water, which can vary with the humidity.

Waste Minimization and Disposal

1. Although cyclohexane has low toxicity, it is not soluble in water and thus cannot be flushed down the drain. Waste cyclohexane should be stored in a glass bottle and eventually given to a hazardous waste contractor. Since cyclohexane is used simply to wash impurities out of the filters, it can be stored and used again in subsequent experiments for the same purpose. This will minimize the waste produced.

2. The amount of filtered particulates is very small and there is no reason to assume they are hazardous waste. In this case they can be put in the trash. Otherwise they should be stored in a closed container and eventually given to a commercial waste disposal facility. There are additional experiments that can be carried out with the particulates, in which case the final disposal is discussed in connection with that experiment.

Data Analysis

1. Determine the average of the initial and final flow rates. Multiply this value by the number of minutes sampled to determine the <u>total volume</u> of air sampled.

2. Calculate the total suspended particulates (TSP) by dividing the mass of collected particulates (in μg) by the total volume of air (in cubic meters).

3. Discuss the difference in the result obtained in the lab and outside the lab.

Supplemental Activity

1. If the outdoors collection was done in a high traffic area, where a good part of the collected particulates are soot, there may be a measurable quantity of PAHs present. An excellent extension to this experiment is the extraction and measurements of the PAHs. This may be accomplished by first extracting the PAHs into cyclohexane. Place the filter in a 150 mL beaker followed by exactly 10.00 mL of high-quality cyclohexane. Cover with aluminum foil and place the beaker in an ultrasonic bath and sonicate for 30 minutes on high. If particulates are noticed, filter the solution into a 25 mL volumetric flask, rinsing the beaker and its contents with three very small portions of cyclohexane, adding the rinsings to the funnel. Fill to the mark.

The solution of PAHs can now be measured fluorimetrically according to the procedure in Experiment 15. If this is done, report the anthracene equivalent of the PAHs present /gram of particulate.

2. A second method of analysis is gas chromatography/mass spectrometry. If this is done, evaporate the sample down to about 2 mL and then analyze on a nonpolar column, such as a DB-5. Report the PAHs found in the analysis. This should only be done if the student has the necessary background in instrumental analysis or organic chemistry.

3. Another interesting extension, or alteration, to the given experiment is to sample the particulate matter in a room filled with cigarette or cigar (or pipe) smoke. Alternatively, a device could be devised to

automatically produce tobacco smoke that could then be collected and analyzed by any of the mentioned methods.

4. It is interesting to observe the particulates collected with a microscope. If a picture of the particulates can be obtained, it would be interesting to compare with pictures of known air particulates to try to identify the particulates.

Questions and Further Thoughts

1. If sample collection does not yield a measurable sample, indicate three ways in which the collected sample mass can be increased.

2. If you took your home air conditioner filter and examined its contents carefully, how would they differ from what you would observe in samples collected (a) outside and (b) in the laboratory?

Notes

1. The flow rates for all pumps should be about equal. If the volumes of air sampled are significantly different, it will be difficult to draw any conclusions based on the darkness of the two filters.

2. When removing the flow meter, pull the meter off the end of the filter holder, being careful that the casing does not come off with the meter and that a good seal is formed between the casing and the tube.

3. A construction site is a good location for sampling.

Literature Cited

1. P. N. Cheremisinoff and A.C. Morresi, *Air Pollution Sampling and Analysis Deskbook,* Ann Arbor Science, Ann Arbor, MI, 1978.

2. R. Delumyea and R. Petel, *Atmospheric Inputs of Phosphorus to Southern Lake Huron,* EPA Report No. 600/3-77-038, Final Report, 1976.

3. A. E.Greenberg, L. S. Clesceri, and A. D. Eaton, Eds., *Standard Methods for the Examination of Water and Wastewater,* 18th ed.,Edition, American Public Health Association, Washington, DC, 1992.

4. S. E. Manahan, *Environmental Chemistry,* 5th Fifth ed., Lewis, Chelsea, MI, 1991.

5. G. M. Masters, *Introduction to Environmental Engineering and Science,* Prentice-Hall, Englewood Cliffs, NJ, 1991.

6. A. V. Nero, Jr., Controlling indoor air pollution, *Scientific American,* 258(5):42-48 (1988).

7. R. W. Shaw, Air pollution by particles, *Scientific American*, 257(5): 96-103 (1987).

Experiment 22

Determination of the Concentration of Carbon Dioxide in the Atmosphere

Objective—The objective of this experiment is to determine the carbon dioxide concentration of the atmosphere by reacting the carbon dioxide with barium hydroxide, precipitating barium carbonate. This experiment also introduces the fundamental ideas of gravimetric analysis, which we studied in an earlier chapter.

Introduction—Although there are about a dozen components of the atmosphere, there are only two major components, oxygen and nitrogen. The other chemical species are present in only trace amounts, as shown in Table 22-1. Of the substances listed, only one has a variable concentration that is important, carbon dioxide. Carbon dioxide is a colorless, odorless gas that, although not toxic, does not support respiration and thus causes suffocation at high concentrations.

There has been an increase in the average temperature of the earth of 0.3–0.6°C over the last century. The substances most responsible for maintaining earth's surface temperature are carbon dioxide and water. Carbon dioxide plays a secondary, but crucial role in maintaining the surface temperature of the earth. The worldwide consumption of fossil fuels (coal, petroleum, and natural gas) has sharply increased the CO_2 level in the atmosphere. A gallon of gasoline (density = 0.70 g/mL) with an approximate composition of C_8H_{18} produces about 8 kg of CO_2, which translates to over 20 billion tons produced annually. Large quantities of carbon dioxide are used for refrigeration and the production of carbonated beverages.

Table 22-1 Composition of Dry Air Near Sea Level

Component	Mole Fraction
Nitrogen	0.78084
Oxygen	0.20948
Argon	0.00934
Carbon dioxide	0.000355
Neon	1.818×10^{-5}
Helium	5.24×10^{-6}
Methane	2×10^{-6}
Krypton	1.14×10^{-6}
Hydrogen	5×10^{-7}
Nitrous oxide	5×10^{-7}
Xenon	8.7×10^{-8}

Scientists have monitored carbon dioxide levels in the atmosphere since 1958. After the last ice age, the level remained fairly constant until the start of the Industrial Revolution, about 300 years ago. Since that time the CO_2 concentration has increased by 25% and the present level in the atmosphere is about 355 ppm.

Approximately one-half of anthropogenic carbon dioxide emissions find a sink in seawater. Large amounts are also removed from the atmosphere each spring and summer because of photosynthesis in plants:

$$CO_2\,(g) + H_2O\,(l) \longrightarrow O_2\,(g) + \text{polymeric}\,(CH_2O)\ \text{(plant fiber)}$$

However, biological decay of plant material occurs in the fall and winter and replaces much of the carbon dioxide.

Theory—Carbon dioxide readily reacts with base to form bicarbonate or carbonate, according to the equations,

$$CO_2\,(g) + OH^-\,(aq) \longrightarrow HCO_3^-\,(aq)$$

and

$$CO_2\,(g) + 2\,OH^-\,(aq) \longrightarrow CO_3^{2-}\,(aq) + H_2O\,(l)$$

If sodium or potassium hydroxides are used, the resulting sodium or potassium bicarbonate or carbonate salts are soluble in water. On the other hand, if barium hydroxide is used, a precipitate of carbonate results,

$$Ba^{2+}(aq) + 2\,OH^-(aq) + CO_2\,(g) \longrightarrow BaCO_3\,(s) + H_2O\,(l)$$

As barium hydroxide is consumed, the pH of the solution and its conductivity decrease considerably. The amount of hydroxide neutralized can be estimated by the conductivity change, but can be more accurately determined by titrating excess OH^- with standard acid after filtering off $BaCO_3$. Filtration is necessary because $BaCO_3$ also reacts with acid.

In this experiment, $BaCO_3$ will be measured gravimetrically since it is the most direct method, is very accurate, and introduces the principles of this important area.

Although usually extremely accurate, a gravimetric method is time-consuming and requires good technique, as well as strict attention to details. The following general principles are necessary to ensure good results:

1. There must be a quantitative precipitation of the precipitate, at least 99.9% complete. This can be ensured if the K_{sp} of the precipitate is small enough, and if not, using excess precipitating agent to shift the equilibrium to the right.

2. It is often necessary to prevent the precipitate from forming a colloid (sol) that will pass through a filter. Heating and stirring the precipitate aids coagulation and the subsequent growth of large crystals. Rinsing the precipitate with electrolyte solutions prevents the precipitate from spontaneously becoming colloidal again.

3. Prevent the coprecipitation of excess reactant ions.

4. Use a final rinse of the precipitate that will replace nonvolatile adsorbed ions with ions that will decompose or volatize at the temperature of the drying oven, preferably between 110 and 200°C.

To see if $BaCO_3$ is precipitated quantitatively, consider the K_{sp}, which at 25°C is 4.9 x 10^{-9}. In a saturated solution its solubility is 7.0 x 10^{-5} mol/L, or 9.6 mg/L. Thus, if 1 g of barium were precipitated out in 100 mL of solution, there would be an error of about 0.1%. When an excess of barium ion is used in the analysis, the reaction will be quantitative.

To prevent the precipitate from reverting to the colloidal state, it is rinsed with an electrolyte solution, in this case, barium nitrate. The barium salt is used to prevent solubility losses of the precipitate. A final rinse with ammonium nitrate replaces excess barium on the surface of the precipitate with the ammonium ion, which is lost on drying. On drying the following reaction occurs,

$$2\ NH_4NO_3\,(s) \longrightarrow 2\ N_2O\,(g) + 4\ H_2O(g)$$

Air Sampling

Most air samplers contain the components shown in the air-sampling train of Figure 22-1.

Figure 22-1 General Diagram of an Air-Sampling System

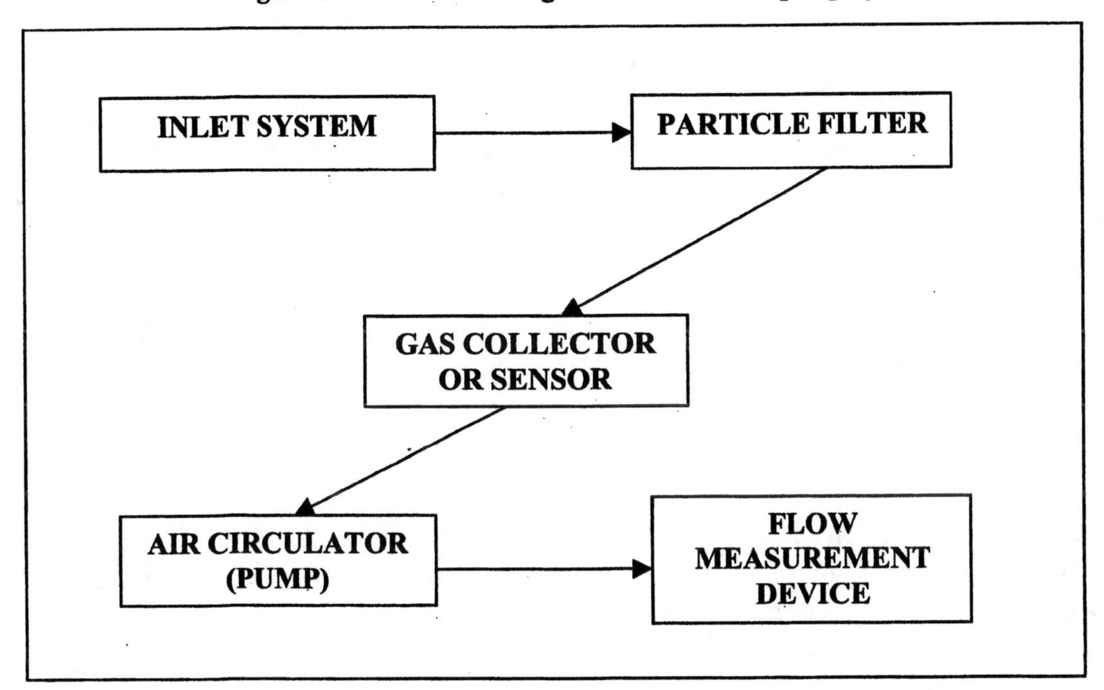

The inlet system must be inert to the analyte being studied and is commonly made of Teflon, glass, or stainless steel. The inlet must be located away from any solid surfaces or physical obstructions. It must also be located away from possible point sources (factories, electric plants, highways, etc.) if the sample is to be representative.

If the sampling objective is to collect airborne particles (an aerosol), the nature of the filter material is important. Commonly used filter materials are glass fibers, cellulose ester, and Teflon membranes. Typical pore sizes for membrane filters are about 0.5 μm.

If the analyte of interest is a gas, a gas collector or sensor follows the particle filter. The collector may be a wash bottle containing a liquid reagent, or it may be an absorbent porous polymer, depending on what is to be collected. For example, SO_2 may be collected in dilute hydrogen peroxide solution where it forms sulfuric acid, which may be analyzed by titration of acidity, measurement of SO_4^{2-}, or solution conductivity:

$$SO_2\,(g) + H_2O_2\,(aq) \longrightarrow H_2SO_4\,(aq)$$

Air is drawn through the system by a pump, which is always used to draw, rather than blow air. This prevents sample contamination from the pump that might occur if the air was drawn through the pump. Flow rates vary for different pumps but typically lie in the range of 0.5–20 L/min for "low-volume" samples, and are about 1–2 m^3/min for "high-volume" samples. The flow rate can be measured with a gas meter, which measures an integrated air volume. Alternatively, a flow-rate meter, such as a rotameter, can be used to measure flow rate, which may be converted to a volume of air by multiplying by the sampling time. Some commercially available gas analyzers have all the components built into a compact arrangement and may have a continuous sensor to measure the concentration of analyte(s).

Safety Issues

1. Safety glasses must be worn at all times in the chemistry laboratory.

2. Use gloves when handling barium hydroxide and its solutions. If any hydroxide gets on your skin, rinse for several minutes with cool water.

Procedure

1. Assemble the sampling apparatus in the order shown in Figure 22-1. In this experiment, the gas collector is a bubbler sampler like the one shown in Figure 22-2. If more than one set-up is available, it would be informative and interesting to locate them at different sites to see if there are differences in the CO_2 levels. Outdoor samples near traffic, for example, may yield different results than those found in the laboratory. If few setups are available, the experiment can be carried out over several lab periods to obtain results over a time span. The pump is a compressor with an adjustable air flow. However, the system is operated by connecting the air input of the pump to the side arm of the bubbler. This draws air into the bubbler and the solution. The air drawn through the system is measured with a flow meter.

Figure 22-2 Bubbler Sampler for Collection of Carbon Dioxide

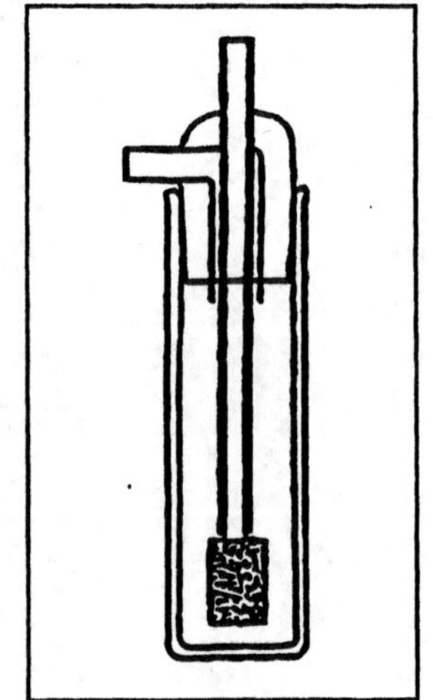

2. Accurately weigh (to 0.1 mg) about 5 g of barium hydroxide into a small beaker or a weighing boat. Quantitatively transfer to the collection tube and add about 150 mL DI water.

3. Attach the tubing from the flowmeter and measure the initial flow rate. It should be about 2 L/min. Collect a sample for 90–120 minutes; measure the flow rate at the end of this time.

4. While the carbon dioxide is being collected, obtain a clean filter crucible of either medium or fine

porosity. Put into a drying oven at 110°C for 0.5 hour, and then transfer to a desiccator to cool. Then weigh (to 0.1 mg).

5. Before and after sample collection, measure ambient temperature and barometric pressure.

6. Disassemble the apparatus, add 75 mL DI water to the tube, mix and then allow the precipitate to settle for 10 minutes. Decant the supernatant liquid quickly through the filter crucible using suction. Then transfer the precipitate to the filter crucible, using 0.1 M barium nitrate wash solution to rinse the collection tube and the precipitate.

7. Finally, rinse the precipitate with several small portions of 0.1 M NH_4NO_3.

8. Suction for an additional minute after all liquid has passed through the funnel. Put the crucible in a drying oven at 110–120°C for 1 hour.

9. Transfer the crucible to a desiccator to cool, then weigh and obtain the mass of $BaCO_3$ by difference.

10. Return the crucible to the oven for 0.5 hour and repeat step 9 until the crucible with sample reaches constant weight.

Waste Minimization and Disposal

1. Excess barium hydroxide solution and barium nitrate solution can be treated with dilute sulfuric acid. This neutralizes the base and also forms a very insoluble precipitate of barium sulfate. The barium sulfate precipitate and the precipitate of barium carbonate can be combined in a waste bottle. These precipitates take little space and can be stored until a sufficient amount is accumulated, at which time it can be given to a commercial waste disposal facility.

2. Ammonium nitrate solutions in small amount and concentration can be disposed of by rinsing down the laboratory drain (after approval by the lab instructor).

Data Analysis

1. Report the mass of $Ba(OH)_2$ used, the collection time, and the mass of $BaCO_3$ produced. From the latter, calculate the mass and number of moles of carbon dioxide collected.

2. Calculate the moles of air collected, using the ideal gas law, $n = PV/RT$. Use the average atmospheric pressure and temperature during the collection period. The volume is the flow rate times the time. Be careful to use the correct units.

3. Calculate the ppm CO_2, which is equal to its mole fraction, and is equal to $\boldsymbol{n}(CO_2)$ / $\boldsymbol{n}$(air), and compare with:

   a. the value presently known for the atmosphere

b. values obtained by other groups in your class

4. Discuss factors that would give low results and factors that would give high results in this experiment.

Supplemental Activity

Supplemental Activity—The remaining hydroxide can be analyzed by direct titration with standard acid to check the results obtained gravimetrically.

Questions and Further Thoughts

1. When carbon dioxide is produced by the combustion of a fossil fuel, water, another greenhouse gas, is produced. Compare the amount of water produced when 1 gallon of C_8H_{18} is burned (see the Introduction) to the amount of carbon dioxide produced. Is the water produced as much of a problem as the carbon dioxide? Why?

2. Are there other pollutants that can interfere with the analysis for carbon dioxide? If so, write the equations for the reactions.

3. Can the carbon dioxide level in the atmosphere be determined by a precipitation titration of the excess barium present in the reaction mixture? If so, how could it be done?

4. Gas chromatography can also be used to determine the CO_2 level in the atmosphere. What are some of the advantages and some of the disadvantages of this method for such an analysis?

Notes

1. A sampling train consisting of a tube containing $Mg(ClO_4)_2$ to absorb water connected to a tube containing NaOH can be used to determine ppm CO_2 in the atmosphere. A difficulty with this method is that when humidity is high, excess water may break through the $Mg(ClO_4)_2$ tube into the NaOH tube.

2. After collecting the carbon dioxide, water is added to the reaction mixture to dilute the hydroxide since sintered glass is very soluble in strong base. Also, the time the base is in the funnel should be kept as short as possible. Without diluting the base the loss is still small, typically being 0.01 g.

3. An aquarium pump can put out more than 3 L/min and could be used in this experiment.

4. Too much frothing occurs if the air flow is greater than 3 L/min.

Literature Cited

1. R. M. Harrison, S. J. de Mora, S. Rapsomanikis, and W. R. Johnston, *Introductory Chemistry for the Environmental Sciences,* Cambridge University Press, Cambridge, England, 1993.

2. G. M. Masters, *Introduction to Environmental Engineering and Science,* Prentice-Hall, Englewood Cliffs, NJ, 1991.

3. D. G. Peters, J..M. Hayes, and G. M. Hieftje, *A Brief Introduction to Modern Chemical Analysis,* W. B. Saunders, Philadelphia, PA, 1976.

4. T. G. Spiro and W. M. Stigliani, *Chemistry of the Environment,* Prentice-Hall, Upper Saddle River, NJ, 1996.

Experiment 23

Collection and Chemistry of Acid Rain

Objectives—The objectives of this experiment are to learn the proper collection and storage technique for rainwater and to analyze a rainwater sample using basic analytical methods for metals and nonmetals.

Introduction—Water has been said to be a "universal solvent," and this is true to a degree. Water is an excellent solvent for many inorganic compounds, but is not as good for most organic compounds. However, even in the case of inorganic substances, there is limited solubility for many molecular and ionic compounds. Minerals persist in nature due to their limited solubility. Thus, the only stage in the hydrologic cycle where pure water is found is at the beginning of the cycle. However, even raindrops contain impurities since the formation of raindrops is nucleated by particles, especially those containing nitrates and sulfates. Rainwater can be an excellent system for testing the atmosphere for various water-soluble pollutants.

Theory—At 25°C the pH of pure water is 7.0. However, as we saw in Experiment 3, rainwater in contact with atmospheric carbon dioxide should have a pH of 5.6–5.7. However, even in remote areas, uncontaminated by either industrial emissions or calcareous dust, rainwater usually has a pH closer to 5.0 since it contains small amounts of both weak and strong acids of natural origin.

The two acids most responsible for acid rain are nitric and sulfuric acids. Both acids can be formed naturally; lightning produces NO_x and volcanoes produce SO_2. Sulfur oxides can also be produced naturally by bacterial decomposition. The nitrogen and sulfur oxides combine with water vapor to form the nitrogen and sulfur acids. The formula NO_x stands for both NO and NO_2. When nitric oxide is formed, it is rapidly oxidized in the atmosphere to NO_2, which then combines with atmospheric water to produce nitric acid. Sulfur dioxide itself can form an acid in water, sulfurous acid, H_2SO_3. However, in the presence of large amounts of oxygen in the atmosphere, SO_2 is catalytically oxidized to SO_3, which reacts with water to form sulfuric acid, H_2SO_4. The reaction is facilitated by UV radiation and is catalyzed by dust particles, the surfaces of buildings, and the metals and carbon in fly ash. Ozone also participates in the oxidation reaction. Sulfur dioxide and nitrogen oxide emissions are cleared from the atmosphere in a few days, eventually being deposited on soil or water either directly by dry deposition in aerosols or indirectly by wet deposition in rainfall.

Globally, anthropogenic emissions of sulfur are comparable in magnitude to emissions from natural sources, but in polluted urban areas more than 90% of atmospheric sulfur is anthropogenic. Coal-fired power plants can produce copious amounts of SO_2.

Nitrogen oxides are produced in the internal combustion engine. The reaction,

$$N_2 + O_2 \longrightarrow 2\,NO$$

is endothermic and is not spontaneous at 25°C. However, in the cylinder of an automobile engine the temperature can be as high as 2400 K during the combustion cycle. At this temperature the equilibrium is shifted to the point where appreciable amounts of NO_x are formed. Before pollution control devices, typical emission levels of NO_x were about 4 g/mi, but with such devices the present auto emission standard for NO_x is less than 0.4 g/mi. Although the amount of NO_x produced per mile is small, the widespread use of the automobile produces huge quantities of NO_x worldwide.

Atmospheric nitric and sulfuric acids do not significantly affect the pH of rainwater unless they are produced in large quantity in urban areas from anthropogenic activities. In polluted areas their concentrations can be quite high and reduce the pH of rain to the point where it is corrosive to metals and construction materials. When this is the case, the rain is called "acid rain" and the pH can be less than 4.0.

It is interesting to note that the pH of rainwater changes during the course of a rainfall. At the start, the concentration of sulfuric acid is the greatest and rain will have its lowest pH. Values as low as pH 3 have been measured.

Soils may neutralize the acids in acid rain, but when their buffer capacity is decreased, soil releases the acids to surrounding bodies of water. Here the decreased pH may have deleterious effects on the plant and animal life.

Stone monuments and buildings most susceptible to acid rain are those made of calcareous materials, including marble, limestone, and calcareous sandstone. Chemically, marble and limestone are identical, but morphologically they differ in crystal size and porosity; limestone is more porous due to its smaller crystal size. Calcareous sandstone is made up of quartz cemented together by calcareous materials. The carbonates present in these materials are subject to dissolution by mineral acids.

Safety Issues

1. Safety glasses must be worn at all times in the chemistry laboratory.

2. Review the safety precautions in Experiment 10 concerning the use of the atomic absorption spectrometer.

Procedure

A: Collection of Rainwater

1. A simple rain collector can be made from two similar bottles. The bottom of a gallon Nalgene bottle is removed with a knife or hacksaw. The open bottle is inverted onto a second (whole) gallon bottle. The connection is sealed with several layers of ParaFilm and finally wrapped in electrical tape. Before assembly, wash the two container parts with a non-phosphate detergent. Rinse thoroughly with tap water, then triple rinse with DI water and drain dry. A collector of this type works well for inorganic analytes, but would not be suitable for organic analytes due to possible contamination from, or interaction with, the sample container or sealants. An even simpler collector is shown in Figure 23-1.

Figure 23-1 A simple Rain Collector

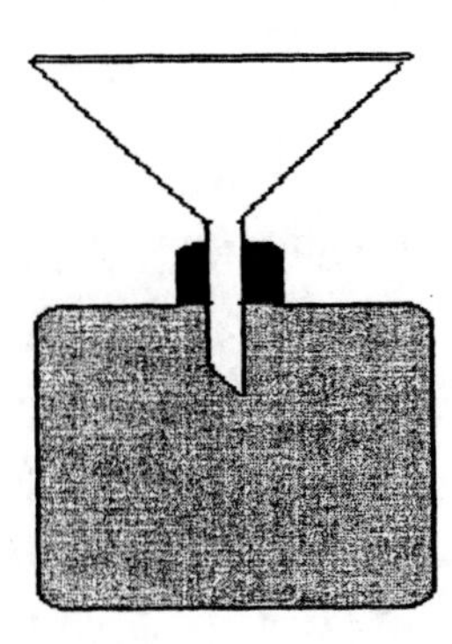

2. Obviously one cannot collect rainwater at any desired time. Collect a sample by planning well in advance. Keep up with weather forecasts to decide when to put the rain collector out. Place the collector in an open area away from buildings, trees, and other obstructions to prevent sample contamination. Note the time collection was started and the time when it was stopped. As with all sample-collection events, make note of any significant activities or physical features in the sampling area, such as proximity to manufacturing facilities, power plants, highways, and other possible sources of pollution, man-made or of natural origin. Wind direction is also an important factor to record. This type of additional information is very beneficial in explaining sample results. When multiple sampling sites are used, especially for extended periods of time, this inventory of the surrounding areas and meteorological conditions over the sampling periods aids the environmental scientist in analyzing patterns and trends of pollution.

B: Analysis of Rainwater

1. After thoroughly mixing the rainwater sample, transfer a portion to a small beaker. Measure and record the pH and temperature of the sample. The best sampling protocols require that primary pH and temperature measurements be made in the field when possible.

2. Use another aliquot of sample to measure specific conductance and temperature of the rainwater.

3. Use the procedures described in Experiments 5 and 7 to determine the sodium and chloride ion concentrations in the rainwater sample using ion-selective electrodes. Standards from those experiments will be available for these measurements.

4. Use the atomic absorption spectrometry methods described in Experiment 10 to determine the sodium ion concentration in the sample of rainwater.

Waste Minimization and Disposal

1. The only waste generated is extra rainwater and sodium chloride standards used for atomic absorption and conductance and ion-selective electrode measurements. None of these are hazardous wastes, and, if necessary, they can be flushed down the laboratory drain.

2. The rainwater can be used for other experiments, and the standards can be stored for future use.

Data Analysis

1. Report the pH of your rainwater sample and compare this value with the expected values for water saturated with CO_2, the values generally found in urban areas, and the value usually found in unpolluted rural areas. (Refer to Experiments 3 and 4.)

2. Use the specific conductance of the rainwater sample to estimate the total dissolved solids in your sample of rainwater. Compare with the dissolved solids in environmental waters from Experiment 6.

3. Submit the calibration curves used to determine the sodium and chloride ion concentrations with ion-selective electrodes. Report the concentrations of both ions and discuss the values.

4. Submit the calibration plot obtained for a series of standard sodium solutions measured by atomic absorption. Report the sodium ion concentration found for the rainwater determined by atomic absorption analysis and compare with the value obtained using an ion-selective sodium electrode.

5. From your analysis of rainwater, what can you conclude about the air quality in your vicinity?

Supplemental Activity

1. A study referred to in the Literature section (Ref. 4) indicates that the atmosphere is a major source of phosphates in certain areas. Standard methods for water analysis can be used to detect phosphate at very low concentrations, and this would be a possible substitution or addition to the present experiment. When analyzing for phosphates, it is imperative that all sampling and testing equipment be free of phosphates. Many detergents are sources of phosphates, and care must be taken to use one that is phosphate-free.

2. If rain samples for a long rainfall event can be collected over a period of time, the variation of pH and specific conductance over time can be determined. The same is true for the metals and chloride ion concentrations.

3. Similar studies can also be done for snow and other forms of precipitation with some modifications to the sampling system.

4. The density and refractive index can readily be determined for rainwater samples and may be used to supplement the previous measurements.

Questions and Further Thoughts—If rainfall is to be analyzed for metals at a time other than a day or two after collection, that sample should be treated with high-purity nitric acid (about 1 ml/L). This affects the pH and conductivity measurements, which should be made as soon as possible after sample collection and before preservation.

Notes—The same standard sodium chloride solutions used to calibrate the ion-selective sodium electrode and chloride electrode can be used to prepare the calibration plot for atomic absorption analysis.

Literature Cited

1. P. A. Baedecker and M. M. Reddy, The erosion of carbonate stone by acid rain, *J. Chem. Ed., 70*(2):104–108 (1993).

2. A. E. Charola, Acid rain effects on stone monuments, *J. Chem. Ed., 64*(5):436–437 (1987).

3. P. N. Cheremisoff and A. C. Morresi, *Air Pollution Sampling and Analysis Deskbook,* Ann Arbor Science, Ann Arbor, MI, 1978.

4. R. Delumyea and R. Petel, *Atmospheric Inputs of Phosphorus to Southern Lake Huron, April-October 1975,* EPA-600/3-77-038, final report, 1976.

5. A. E. Greenberg, L. S. Clesceri, and A. D. Eaton, Eds, *Standard Methods for the Examination of Water and Wastewater,* 18th ed., American Public Health Association, Washington, DC, 1992.

6. D. W. Schindler, Effects of acid rain on freshwater ecosystems, *Science, 239*:149–153(1988).

GLOSSARY

A

Absorbance. A quantitative measure of the decrease in incident light passing through a sample. Mathematically, it is defined by $A = .\log(P_0 / P,)$ where P_0 is the radiant power of the initial light beam and P is its radiant power when leaving the solution.

Accuracy. How near the result of an analysis is to the actual, or true, value.

Acid Digestion. The treatment of a sample with acid to free metal analytes into solution for analysis.

Acid Extractables. Organic analytes that are removed from acidified water with dichloromethane.

Acidity. The capacity of a system to absorb or neutralize hydroxide, OH^-.

Acid Rain. Rain with an acid content due mainly to sulfur and nitrogen oxides, SO_x and NO_x. The acids that contribute most to acid rain are sulfuric and nitric acids, H_2SO_4 and HNO_3. The pH of natural, unpolluted rain is about 5.6. Only rain that is significantly more acidic than this is considered to be "acid rain," the pH of which is commonly less than 5.0.

Activity. In solution it is the effective concentration of a solute. When a solution becomes dilute, the activity of a solute is close to the actual concentration. At the ppm level, the activity of a solute is practically identical to its concentration.

Activity Coefficient. A parameter that indicates how the actual concentration of a solute is related to its apparent concentration. Its defining relationship is $a_i = \gamma_i[\,i\,]$, where a_i is the activity and $[\,i\,]$ the molar concentration of i and γ_i is a dimensionless quantity called the activity coefficient. The activity coefficient depends on ionic strength.

Alicyclic Hydrocarbons. Saturated hydrocarbons in which carbon atoms form rings. An example is cyclobutadiene. They are present in petroleum and refined products.

CH_2—CH_2
| |
CH_2—CH_2

Aliphatic Hydrocarbons. Saturated hydrocarbons having either straight chains, as in pentane, $CH_3CH_2CH_2CH_2CH_3$, or branched chains, as in 2-methylbutane. It is one of the two major groups of hydrocarbons. They are composed of alkanes, alkenes, and alkynes.

CH_3—$CHCH_2CH_3$
  |
CH_3

Alkalinity. The capacity of a system to absorb (react with) hydronium ions, H_3O^+.

Aliquot. A measured portion of a sample that is taken for analysis.

Ambient. Surrounding or usual, as in ambient temperature, or ambient atmospheric pressure.

Analyte. The species the analyst wishes to measure or detect.

Analytical Balance. A balance that can weigh with a precision of at least one part in 10^5 at its maximum capacity. It is capable of weighing to 0.1 mg.

Anthropogenic. Human in origin or a result of human activities. Some hydrocarbons are anthropogenic in origin; that is, they are derived from the refining of petroleum.

Aquifer. Groundwater found in strata of porous rocks that are separated both above and below by impervious layers of rock. It connects to the surface through a "recharge area" into which rainwater and/or surface waters can enter the subterranean layer.

Aromatic Hydrocarbons. Hydrocarbons in which the carbon atoms are bonded in a ring structure, with alternating single and double bonds (that is, conjugated π bonds) and having an enhanced stability due to electron delocalization.

ASTM. American Society for Testing and Materials. An organization that devises standards for materials characterization and use.

Atomic Absorption (AA). An instrumental method of quantitative analysis based on aspirating a metal sample in solution into a flame where elements are atomizied, followed by monitoring the absorption by the atoms of specific wavelengths of light passed through the flame.

B

Background Correction. A procedure used to correct for background noise adding to the instrument signal when determining trace amounts of elements.

Background Level (or Concentration). The average concentration of an analyte present in the environment of a geographical area due solely to natural processes.

Base-Neutral Extractables. Compounds extractable from basic or neutral aqueous solutions with organic solvents.

Basicity. The ability of an aqueous solution to neutralize acid due to its having a pH greater than 7.00.

Beer-Lambert Law. Mathematically $A_i = \varepsilon_i b[\,i\,]$ or $A_x = a_x b C_x$ where A is absorbance, $[\,i\,]$ is molar concentration, C_x is any other concentration unit, ε_i is the molar absorptivity (or extinction coefficient), and a_x is the absorptivity.

Bias. The agreement of a measurement (or an average of several measurements) of a quantity with an accepted, or true value. A consistent deviation of measured values from the actual values and due to a systematic error the method; or caused by some artifact or quirk of the measurement process. Sample contamination or extraction inefficiency are two possible sources of bias.

Bioaccumulation. The increase in the amounts of certain substances in biological systems due to selective partitioning. It is a phenomenon that occurs in biological systems that are exposed to various substances in their environment. The substances are concentrated in organisms by various methods of uptake and are absorbed faster than they are metabolized or excreted. If at each level in the food chain these bioaccumulated substances increase in concentration, the process is called **biomagnification**.

Biochemical Oxygen Demand (BOD). A measure of water quality quantitatively equal to the decrease in the dissolved oxygen content per liter of a sample over a 5-day period and at a temperature of 25°C. Its units are mg/L.

Bioconcentration Factor (BCF). The ratio of the concentration of a pollutant in an organism to its concentration in the surroundings. It indicates the degree of bioaccumulation for a particular compound.

Biodegradable. Capable of being broken down by biological processes into simpler inorganic and biological substances.

Biogenic. Originating from a natural, or biological source. For example, biogenic hydrocarbons include alpha-pinene and isoprene from trees.

Biomagnification The increase in concentration of a pollutant along the food chain.

Blank A reference system, usually a solution, commonly used in many analytical methods. It contains everything that is used to treat the sample but does not contain the analyte. There are many kinds of blank, several of which are given in this glossary.

Blank Analysis. A quality-control procedure that tests for contamination of analyte in the reagents used for an analysis.

Buffer. A system that resists a pH change when small amounts of strong acid or strong base are added to the system. They are usually composed of a weak acid and a salt of its conjugate base, or a weak base and a salt of its conjugate acid.

Buffer Capacity. The amount of strong acid or base needed to change the pH of a buffer system by one pH unit. It is an <u>extensive</u> factor, whereas pH is an <u>intensive</u> factor.

C

Calibration Blank. The calibration blank is used to determine the null (zero) reading for the instrument response when determining the concentration calibration curve. It is ordinarily a solution that is as free of analyte as possible, and has the same volume of the same solvent used for the preparation of the calibration standards.

Calibration Curve. A plot of instrumental response versus the amount of analyte in standards. It is usually a straight-line result.

Calibration Standards. A series of standard solutions that are used to calibrate an instrument (i.e., prepare the analytical curve).

Carbonaceous Biochemical Oxygen Demand (CBOD). The amount of oxygen needed to oxidize any carbon-containing matter present in a water sample.

Carcinogen. A substance that can cause cancer at various sites in an organism.

Chemical Oxygen Demand (COD). A quick, convenient chemical method for approximating the BOD of a sample of water. In one method, an acidic solution of potassium dichromate is used to oxidize organic matter and the amount of dichromate remaining is determined either by titration or colorimetry. The oxidation can be catalyzed by silver ion, and mercuric sulfate avoids the interference due to the presence of chloride ion. The COD is equal to the equivalent amount of oxygen needed to oxidize the organics present by means of a strong chemical oxidizing agent. In this method substances are oxidized that ordinarily would not be oxidized in a BOD test. For this reason, the COD is usually greater than the BOD.

Chlorinated Hydrocarbons. Hydrocarbons in which some or all the hydrogens have been replaced by chlorine atoms. 1,1,1-Trichloroethane, $\mathbf{CH_3CCl_3}$, is an example.

Chlorofluorocarbons (CFCs). A family of inert, relatively nontoxic, usually nonflammable, and easily liquefied compounds that are used in refrigeration, air conditioning, packaging, insulation, and aerosol propellants. Since CFCs are not readily degraded in the lower atmosphere, they migrate to the upper atmosphere where they decompose in UV radiation, forming chlorine atoms that destroy ozone.

Clean-Up Methods. Procedures used to eliminate interferences in sample extracts prior to analysis.

Colorimetry. A quantitative analytical method applicable to most metals and many nonmetals as well as organic compounds. It is based on the Beer-Lambert law and requires an electronic absorption of the analyte (or a derivative) in the visible region of the spectrum. The intensity of color indicates the concentration of a particular substance.

Contaminant. A substance that changes the normal composition of the environment. Contaminants become pollutants when they have an adverse effect on the environment.

Contamination. The presence of an undesirable substance in a sample being examined. A substance inadvertently added to the sample during sampling and/or the transportation-storage-analysis procedure.

Control. A sample used to help judge the results of a procedure.

Correlation Coefficient (*r*). A regression coefficient having values between –1 and +1. Its sign depends on the sign of the slope. A value of exactly 1 in absolute magnitude indicates a perfect straight line or perfect correlation of the points to the equation.

Critical Micelle Concentration (CMC). The concentration of surfactant at which micelles begin to form. At this concentration, detergent action begins.

D

Density. An intensive property of an element, compound or mixture, it is equal to the mass divided by the volume (**density = mass / volume**). For a pure substance, the density is an important characteristic of that substance and is often reported together with its melting and boiling points.

Detection Limit. The lowest concentration of a chemical species that is statistically significant in a method of analysis or a particular instrument. It is also the smallest concentration that can be clearly differentiated from zero. Quantitatively, it is the minimum concentration of a substance that produces a signal-to-noise ratio of 2.

Detergent. A cleaning agent that contains surfactants which reduce the surface tension of water.

Deviation The difference between a measured value and the mean value for an analysis. It is a measure of precision.

Deviation, Average. The sum of all the deviations divided by the total number of results (of the analysis).

Deviation, Standard. A measure of the precision of an analysis that has a statistical basis. It is given by the equation $s = [\Sigma d_i^2/(N - 1)]^{1/2}$, where d_i is the deviation of an analysis and N is the number of results.

Dispersion. The three-dimensional spreading of a pollutant. This can be approximated using "models" which seek to define how the concentration spreads with distance from the source.

Dissolved Oxygen (DO). The mass (in mg) of oxygen dissolved in water per liter of solution when expressed in parts per million. It is a measure of water quality and is equal to 8.7 mg/L in pure water at 25°C and 1 atm pressure, or 14.7 mg/L at 0° and 1 atm pressure.

Dissolved Solids or Total Dissolved Solids (TDS). A measure of water quality, equal to the total mass of dissolved solids per liter of solution (usually expressed in parts per million). These solids cannot be separated by filtration.

Dry Deposition. The gravitational settling of particles and the transfer of trace gases and aerosols from the atmosphere to the surface of the earth.

Dry Weight. The mass of a sample based on the percentage of solids. It is the mass of a sample after drying it in an oven.

E

Ecology. The study of living organisms and their relationship with one another and their environment.

Ecosystem. An organism or group of organisms, and their surroundings.

Effluent. The fluid (gas or liquid) that flows from a system, process, or receptacle into surface waters or into the atmosphere.

End Point. The point in a titration when the indicator changes color, indicating that the volume of the titrant is about equal to the equivalence point volume.

Environmental Chemistry. The branch of chemistry that studies the chemical properties and changes that occur in the environment.

Equipment Blank. A type of field blank used to determine contaminants that might be introduced by the sampling device. A solution that is as free of analyte as possible is opened at the sampling site, poured over or through the sample-collection device, collected in a sample container, and returned to the laboratory for analysis.

Equivalence Point. The point in a titration when an equivalent amount of titrant has been added to the analyte. For example, if 2.70 mmol of hydrogen ion is titrated with standard base, it is that volume of titrant that contains 2.70 mmol of hydroxide.

Error. The difference between the result of an analysis and the true value.

Error, Relative. The error divided by the true value—often multiplied by 100 or 1000 to obtain the relative error in percent or parts per thousand (ppt), respectively.

Estuary. An arm of the ocean into which fresh water from a river mixes with seawater. Estuaries have a salinity that is high compared to freshwater. Silting is often heavy in estuaries due to the neutralization of suspended colloidal material, allowing it to grow in size and settle out. Estuaries provide good nursery and breeding grounds for fish.

F

Field Blank. A blank used to determine which contaminants might be introduced into the sample during sample collection, transport, and storage. It is usually a solution that is as free of analyte as possible that is transferred from one vessel to another at the sampling site.

Flash Point (FP). The flammability of a substance is determined by measuring its flash point. A waste is hazardous if it is flammable, and it is flammable if it has a flash point less than 60°C (140°F). The test consists of heating the liquid and periodically exposing it to a flame until the mixture of vapor and air ignites at the surface. The lowest temperature at which a flammable liquid provides sufficient vapor to form an ignitable mixture with air is the flash point. After ignition, combustion does not continue.

Fly Ash. Particulate material, consisting of inorganic matter, that results from the burning of coal. It consists of particles of a size that causes it to become entrapped in the lungs of humans. The main constituents are oxides of aluminum, calcium, iron, and silicon. Other elements present include magnesium, sulfur, titanium, phosphorus, sodium, and potassium. Carbon (soot, carbon black) is a significant constituent of fly ash.

Fossil Fuels. Flammable, energy-rich substances derived from the long-term decomposition of animals and plants under great pressure. The chief fossil fuels are natural gas, coal, and petroleum.

Freons. Chlorofluorocarbons having one or two carbon atoms with chlorine and fluorine atoms directly bonded to carbon. The freons are very stable and are often gases at room temperature and 1 atm pressure. They are excellent refrigerants and in the past were also used in aerosol cans as propellants due to their lack of chemical reactivity. They have a detrimental effect on the ozone layer, and their use has been sharply curtailed. An example is dichlorodifluoromethane, $\mathbf{CF_2Cl_2}$.

Fume Hood. An enclosure that is exhausted through the rear, keeping fumes or other emissions (e.g., aerosols) generated within it from being breathed by the user.

G

Gaussian Distribution. A distribution of results that occur when many small, independent deviations in measurements tend to cancel one another, yielding an average value that is close to the true value.

Groundwater. Surface water that has migrated downward through soil pores into geological strata. It comprises more than 90% of all fresh water, and 50% of the earth's population uses groundwater as a primary source of drinking water.

H

Halogenated Hydrocarbons. Hydrocarbons in which one or more hydrogen atoms have been replaced by fluorine, chlorine, bromine, and/or iodine. These substances are generally very unreactive and nonflammable.

Hardness (Hard Water). Once defined as the capacity of a water to precipitate soap, hardness is now defined as the property of water that contains metallic ions that react with the large anions of soaps to form a precipitate termed "scum." For practical purposes, it is the expression of the total concentrations of calcium and magnesium ions present in the water, and is usually expressed in terms of mg $CaCO_3$ per liter of solution.

Hazardous Waste. Substances that can pose a threat to the health or safety of the environment or humans. These substances possess at least one of the following four characteristics— flammability, corrosiveness, reactivity, or toxicity— or they appear on special EPA lists.

Heavy Metal. A pollutant metal of high atomic number, or a metal that occurs in the bottom part of the periodic table. These metals usually have a high density and tend to precipitate sulfur-containing enzymes. These pollutants can damage biological systems at low levels and tend to accumulate in the food chain.

Henry's Law. A quantitative expression of the solubility of a gas in a liquid in terms of its partial pressure above the solution. In equation form it is $[\,i\,] = K_H P_i$ where $[\,i\,]$ is the molarity of substance i, P_i is its partial pressure, and K_H is called the Henry's law constant.

Humic Substances. Organic matter that gives soil its characteristic dark color. The color is primarily due to a material called humus derived from photosynthetic plants. It is the residue from the microbiological degradation of vegetation, and thus itself is resistant to further degradation.

These materials are naturally occurring complexing agents. The binding of metal ions, especially iron and aluminum, is one of the most important environmental properties of humic substances. Chemically, they are high molecular weight, polyelectrolytic macromolecules that contain a carbon skeleton with a high degree of aromatic character and many functional groups containing oxygen.

Hydrocarbons (HCs). Organic compounds that contain only hydrogen and carbon. There are several classes of hydrocarbons, including the alkanes, cycloalkanes, alkenes, alkynes, and aromatics. Hydrocarbons are the chief constituents of the fossil fuels.

Hydrolysis. The chemical reaction of a substance with water. Hydrolysis is a major step in the microbial degradation of many pollutant compounds. It frequently occurs with weak acids

$$NH_4^+ + H_2O \rightleftharpoons NH_3 + H_3O^+$$

and weak bases,

$$CH_3COO^- + H_2O \rightleftharpoons CH_3COOH + OH^- .$$

It is an important reaction of small, highly charged metal ions, and thus is significant for heavy metal ions in the environment.

Hydrophilic. Attracted to water. For example, ions and hydrogen-bonding substances interact with water through ion-dipole and hydrogen-bonding interactions, respectively, and thus are attracted to water.

Hydrophobic. Repelled by water and thus having a very low solubility in water. For example, the oils on some birds repel water. Hydrocarbons are hydrophobic.

I

Interferences. Substances that obscure the measurement of analyte by introducing an unrelated instrumental signal or reacting similarly to the analyte.

Internal Standard (IS). Compounds added to standards, blanks, matrix spikes, samples, sample digestates, and sample extracts at a known concentration prior to instrumental analysis. They are used to quantitate target compounds. The IS provides a standard for determining retention times and response factors.

The internal standard is a compound that has similar chemical properties to the compounds of interest, but is not usually found in the environment. Thus, it does not interfere with the detection of the analyte. An internal standard for PAHs is 1,2-diphenylbenzene, a substance not found naturally, but having a retention time that is about the average of those PAHs normally found in samples.

Inversion. A meteorological condition in the troposphere in which there is increasing temperature with increasing altitude, the reverse of normal behavior. An inversion is caused by a warm air mass overriding a cold air mass in the frontal area. Such conditions are stable. An inversion limits the vertical circulation of air, resulting in the localized entrapment of air pollutants.

Ionic Strength. A property of a solution that affects the activity of dissolved ions. It alters what is termed the "matrix," thereby affecting an analysis. It is quantitated through the equation $\mu = \frac{1}{2}\sum C_i Z_i^2$, where Z_i is the charge on the ion and C_i is its concentration.

K

K-Series. A list, developed by the EPA, of hazardous wastes that result from specific industrial processes.

L

Landfill, Sanitary. A dump or repository for nonhazardous solid wastes that are deposited into ditches, spread into layers, compacted, and then filled in. In recent years plastic liners have been used to contain fluid runoff. When the fill is complete, it is capped to avoid surface water contamination.

Landfill, Secure. Disposal sites for hazardous waste. They have containment devices, such as liners and a leachate-collection system to minimize the chances of releasing hazardous substances into the environment.

Landfill Leachate. Liquid that drains from the bottom of the site into the ground.

LAS. Linear alkyl benzene sulfonates. Biodegradable, anionic detergents that have a linear alkyl chain with a benzene sulfonate group attached at any point on the chain except at the ends. They are the most widely used anionic surfactants. They are composed of many related compounds with variable chain lengths and many isomers.

Leachate. When wastes come into contact with water, some pollutants dissolve and become concentrated in the liquid. This complex mixture is called "leachate," and its analysis often yields useful information about the system under study. Leachate is water that has become contaminated by wastes as it percolates through a terrestrial source, usually a waste-disposal site.

Ligand. A Lewis base that can attach to a metal ion (a Lewis acid) to form a coordinate covalent bond, resulting in a complex ion or coordination compound.

Limnology. The scientific study of fresh water ecosystems, including their biological, chemical, and physical characteristics.

Listed Waste. Wastes listed as hazardous by the EPA because the dangers they possess are self-evident.

Local Control Site. An area where a control sample is taken near the time and place where the samples of interest were taken.

M

Matched-Matrix Field Blank. A common type of field blank that is used to detect and quantitate contaminants that may be introduced at any point in the procedure, from sample collection to the final analysis. The matrix is made as similar to the sample matrix as possible.

Material Safety Data Sheet (MSDS). OSHA has established guidelines for the descriptive data that should be provided on a data sheet to serve as the basis for written hazard communication programs. The purpose is to have those who make, distribute, and use hazardous materials responsible for effective communication.
This paper contains all the information, properties, health effects, and environmental risks of the substance.

Matrix. The background environment of a sample that contains a substance under study. It is the predominant material containing the sample to be analyzed. For example, an atomic absorption study of the manganese in seawater must take into account the high background concentrations of other ions present (the matrix).

Mean, Arithmetic. The sum of all results divided by the total number of results; $\bar{X} = \Sigma X_i/N$ where X_i represents a result and N is the total number of results.

Mean, Geometric. The Nth root of the product of each result. Mathematically it is expressed by $X_{geo} = [X_1X_2 \ldots X_N]^{1/N}$ where the X_i are the individual results and N is the total number of results.

Median. The middle value of a set of results when the results are placed in increasing order. If the number of results is even, it is the average of the two middle results.

Method Detection Limit (MDL). The minimum level of a target analyte that can be determined with 99% confidence. It is the concentration of the analyte that when processed through the complete method produces a signal with a 99% probability that it is different from the blank.

Mineral. A natural, inorganic material that has a relatively constant composition. Minerals can have an important effect on their immediate environment. Rock can be a mass of pure material (quartz is pure silicon dioxide) or a composite of two or more minerals (granite is a mixture that usually contains mica, quartz, and feldspar).

Mobile Phase. In chromatography, the moving phase with which partitioning from the stationary phase occurs. In gas chromatography, it is usually a gas (frequently helium). In thin-layer chromatography (TLC) it is a liquid or liquid mixture.

Mutagen. A substance that can alter an organism's DNA, which can ultimately alter the organism itself.

Mutagenesis. Alteration of parental DNA by a foreign substance that results in mutations in offspring.

N

National Pollutant Discharge Elimination System (NPDES). The discharge criteria and permitting system that was established by the EPA to implement the Clean Water Act (CWA). This system requires a permit by a discharger of wastes, often called a permit to pollute.

Nonpoint Source. A pollution source that has no specific location. Stormwater and agricultural runoff are two examples.

***n*-, Normal.** A chemical prefix meaning having a straight chain, or nonbranched structure.

Nutrients. Chemical substances necessary for the growth of plants and animals. Some of the more important nutrients are nitrogen, phosphorus, carbon, and potassium. However, many metals and nonmetals, including sulfur, chlorine, and iodine, are also essential.

O

Octanol-Water Partition Coefficient. A parameter defined quantitatively by $K_{ow} = [\,i\,]_{octanol}/[\,i\,]_{water}$, where $[\,i\,]_{octanol}$ is the molarity of species i in the octanol phase and $[\,i\,]_{water}$ is the molarity of species i in the aqueous phase. It is used to quantitate the tendency of a pollutant to be bioaccumulated by an organism.

Outfall. A waste disposal into a stream or other receiving water. Many industries have outfalls through which industrial wastes are dumped into the channels of rivers. Some industries combine industrial and domestic wastes to dilute pollutants.

Outlier. A sample value that does not agree in magnitude with other sample values.

Oxidizer. A substance that supports combustion. Examples are nitrates and chromates.

P

Particulate Matter (P.M.). Particles in the atmosphere with sizes from 0.5 mm down to molecular dimensions. This constitutes the most visible and obvious form of air pollution. Pollutant particles are frequently found suspended near pollution sources.

Parts-Per-Million (ppm). A concentration term used for very dilute solutions. Mathematically, it is the mass of the analyte divided by the sample mass, times 1,000,000; (analyte mass/sample mass)(1,000,000).

Petrogenic. Hydrocarbons having their origin in crude oil (petroleum). Natural undersea seepages produce hydrocarbons of petrogenic origin.

Petroleum. Crude oil. An oily liquid found naturally in rock formations. It is composed mainly of hydrocarbons but contains small quantities of many metals, nitrogen, phosphorus, and sulfur.

***Ortho*-Phosphate.** An oxyanion having the structure shown. It is the most abundant form of natural

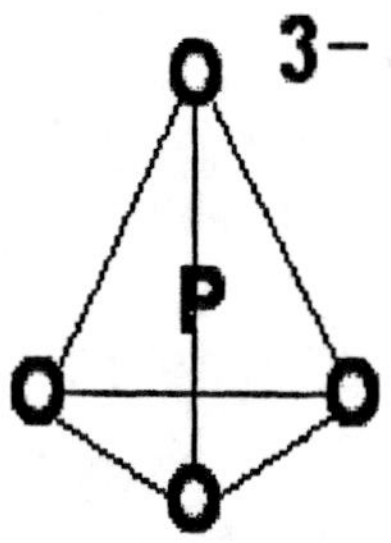

phosphorus. Salts of the phosphate ion are applied to agricultural and residential areas as fertilizers. Phosphate is also a good ligand for complexing metal ions and has been used in the past for water softening. In excess in the environment it causes eutrophication.

Plasticizer. A substance that when added to a polymer improves its characteristics, especially its elasticity. A variety of organic compounds act as plasticizers, including high molecular weight esters.

Plume. When pollutants enter the environment via smoke or water, the stream often remains intact for great distances, keeping the pollutants relatively concentrated. The expanding stream is the plume.

Point Source. A pollution source that is definite, such as a particular industrial operation or a leaking underground gasoline storage tank.

Pollutant. A substance that is detrimental in some respect to the environment.

Polychlorinated Biphenyls (PCBs). Substances having the general molecular formula $C_{12}H_yCl_{10-y}$ (y = 0–9) and the structural formula shown. There are many degrees of

*** Points of Chlorine Substitution**

substitution, with as few as one chlorine per molecule to the fully substituted species with 10 chlorine atoms per molecule. Some isomers are much more carcinogenic than others. PCBs when heated in the presence of oxygen can form the extremely toxic polychlorinated dibenzofurans. PCBs are very stable and are only slowly biodegraded. They have been used as plasticizers, coolant/insulation fluids in transformers and capacitors, and in many other applications. They are universally distributed in the environment.

Polycyclic Aromatic Hydrocarbons (PAHs), or Polynuclear Aromatic Hydrocarbons (PNAs). Aromatic ring systems with two or more fused rings with conjugated π bonds. These substances are frequently produced at high temperatures in oxygen-deficient conditions, especially in incinerators. They are among the most carcinogenic substances known.

Potable Water. Water that contains no harmful or otherwise objectionable impurities and is safe to drink.

Precision. A measure of the agreement of a set of results. There are several measures of precision, some of the more important of which are relative average deviation, relative standard deviation, and relative range.

Preservatives. Substances added to samples to assure the continued presence, or stability, of the target analytes at the same level as when the sample was collected.

P-Series. A list, developed by the EPA, of acutely hazardous wastes originating from commercial chemical products, intermediates, and residues.

pX. A convenient way of expressing very small numbers, usually concentrations. Quantitatively, it is defined by $\mathbf{pX = -log_{10}[\mathit{X}]}$, where $[X]$ is the molarity of X. A decrease in concentration of a factor of 10 increases the pX by 1.00 unit. A more accurate definition uses activity instead of molarity.

Pyrogenic. Hydrocarbons produced at high temperatures, as in forest fires, usually as a result of incomplete combustion. These hydrocarbons can be used to trace the origin of environmental hydrocarbons. Some important hydrocarbons of this type are fluoranthene, phenanthrene, and pyrene.

R

Random or Indeterminate Errors. These are errors that cannot be eliminated. They may result from reading a device, such as a ruler or a buret, and are usually small. If several people read a scale there will be several reported values, and if the errors are truly random, the average of the readings will be close to the exact value. Random errors can be estimated using standard statistical methods.

Range. A measure of precision equal to the difference between the largest result and the smallest result in a group of measurements. Results with a small range are more precise than those with a large range.

Range, Relative. The range divided by the mean, often multiplied by 100 or 1000 to give the relative range in percent or ppt, respectively.

Reagent Blank (Method Blank). The reagent blank is used to detect possible contamination that results from preparing, processing, and analyzing the sample. An analyte-free sample is processed using all the reagents in the procedure and is then analyzed to determine any interferences present.

Receiving Water. The natural water, lake, stream, or ocean into which an outfall flows. It is the water directly impacted by the pollutants from an outfall.

Refractory. Organic compounds that resist decomposition (pyrolysis) when subjected to high temperature are said to be refractory. Such compounds may condense to form even larger molecules, such as the PAHs.

Replicate. Repeated operation occurring within an analytical procedure. Two or more analyses for the same constituent from an extract of a single sample constitutes replicate extract analyses.

Representative Sample. An aliquot of the system being sampled which, when analyzed, gives the same results that the whole system would give if analyzed.

Response Factor. A factor used to convert an instrument response to the mass of analyte.

Retention Time. In chromatography, the time it takes for a particular substance to elute (reach the detector) after being injected into the chromatographic system. It is the time between injection of the sample (in gas chromatography) to its elution at the detector.

Runoff. Water that flows over land to lakes or streams during and after rain.

S

Salinity. A measure of the total dissolved solids in a sample of water. Quantitatively, it is defined as the mass, in grams, of dry salts in 1 kg of water. The average for all oceans is about 35, often expressed as 35‰, meaning parts per thousand.

Saltwater Intrusion. The gradual replacement of freshwater by saltwater in coastal areas, where excessive pumping of groundwater occurs.

Sampling. Choosing and obtaining a representative aliquot of a physical system for testing and/or analysis.

Sediment. The layers of solid material present at the bottom of natural bodies of water. It is composed primarily of mineral and organic matter, the relative amounts of which vary considerably, even at sites close in proximity. Typical sediments consist of mixtures of silt, sand, organic matter, clay, and various other minerals. The relative amount of inorganic and organic matter affects the amounts and types of pollutants present. Sediments are the sink for many chemical substances, especially heavy metals and organic compounds, and many of these can be readily taken up by organisms.

Sedimentation. The settling of material denser than the suspending medium by gravitational force.

Silt. Suspended matter in environmental waters that has a variety of compositions depending on time, location, rainfall, and other factors. It is frequently composed of minerals, tannic and humic substances, detritus from fish, and a variety of other organic matter. It is important in the fate and transport of many types of pollutant and is very absorptive for hydrocarbons and metal ions. When enough material sorbs into silt, it eventually sinks to the surface of the sediment. Being colloidal in nature, it has an electrical charge that can be neutralized by saline water. Thus, in an estuary this results in siltation and the resulting surficial sediment is often very rich in a variety of pollutants if the stream is in an industrial or high-activity area, especially with boating and shipping.

Sink. A long-time repository for a pollutant; it is a location where a pollutant tends to deposit. For example, bone is a sink for lead since lead ion has a size comparable to calcium ion and can replace calcium in bone.

Smog, Photochemical. A combination of smoke and fog laced with SO_2. It is a major air pollution problem and is due mainly to the automobile. For smog to form, relatively stagnant air must be subjected to sunlight at low humidity in the presence of NO_x and hydrocarbons. Tropospheric ozone is the major culprit in smog formation.

Soap. The salt (usually sodium or potassium) of a long-chain fatty (carboxylic) acid (usually C-10 to C-18). A soap is a surfactant and reduces the surface tension of water.

Solid Support. In chromatography the stationary phase is adsorbed onto the surface of the solid support to facilitate its handling and to provide a large surface area.

Spectrometry. Spectroscopic measurements made at one particular wavelength.

Spectrophotometric. Spectroscopic measurements made over a range of wavelengths.

Standard Curve. The curve that results when a plot of the instrument response versus concentrations of known analyte standards is made. It is used to determine the concentration of the analyte from the instrument response for the analyte sample.

Standard Solution. A solution of accurately known concentration, usually known to four or more significant figures. Such solutions are often used to construct a calibration plot.

Stationary Phase. In chromatography, one of the two phases between which partitioning occurs for the components of the mixture being separated. In gas chromatography, it is a solid or a liquid phase.

Stock Solution. A concentrated standard solution of analyte that is diluted to prepare less concentrated standards.

Stratosphere. The layer of the atmosphere above the troposphere. It extends from 12 to 70 km. The temperature of the air increases at increasing altitude.

Surface Active Agent. A substance which, when added in small amounts to water, reduces its surface tension significantly by becoming adsorbed at the air/water interface.

Surface Water. Water found in lakes, rivers, and the oceans. Also artificial impoundments, runoff, and other waters.

Surfactant. Short for surface-active agent.

Suspended Solids. Colloidal particulates. These substances are not filterable due to their small size, usually between 10^{-6} and 10^{-9} m. Also, they do not settle out readily due to bombardment by molecules (Brownian motion).

Systematic Error. An error that usually results in a consistent deviation (bias) in a final result. This kind of error cannot be statistically estimated, but can be eliminated.

System Blank (Instrument Blank). The instrument background, (baseline) response in the absence of a sample. If it is not zero, it's value is subtracted from the instrument response.

Systemic Toxicity. Toxicity in an organism that occurs at an organ remote from the exposed site. The effects are usually concentrated in a particular target organ in the biological system. Chlorinated hydrocarbons absorbed through the skin (or lungs) can cause liver and kidney damage, for example.

T

Teratogen. A chemical substance that alters parental DNA, resulting in birth defects.

Total Organic Carbon (TOC). A measure of water quality used to characterize the total of dissolved and suspended organic matter in a sample of water. It is an expression of the total organic content of a water sample from which the BOD can be estimated. It is frequently measured by catalytically oxidizing carbon to carbon dioxide, which is then measured. This can be done readily and accurately using an instrumental method. Values for groundwater are about 1 mg of carbon per liter of water.

Toxicity Characteristic Leaching Procedure (TCLP). A test developed by the EPA to determine if a waste is hazardous by measuring the potential of the material to release into the environment where it could cause harm to organisms. The test simulates the leaching conditions found in landfills and identifies substances that will leach toxic materials if placed in a landfill. In the test a waste is extracted for 24 hours with a solution of acetic acid. The extraction is simply the dissolving of any constituents that are soluble in the acetic acid solution. The residual acid solution, or leachate, not the solid waste, is analyzed for several substances given in the EPA procedure. It is assumed that if acetic acid does not solubilize the contaminants, neither will the action of rainwater percolating through the landfill.

Toxins. Substances that can harm an organism by biochemical means.

Transducer. A device that transforms information from the chemical domain into the electrical domain. Typical examples are conductivity cells and ion-selective electrodes.

Trihalomethanes (THMs). Chloroform, $CHCl_3$, and dibromochloromethane, $CHBr_2Cl$. They are suspected carcinogens formed from humic substances during the disinfection of raw municipal drinking water by chlorination.

Trip Blank. This blank is used to detect any contamination between the collection site and the lab. Contamination can be caused by sampling equipment. The blank consists of the sampling media used for sample collection.

U

U-Series. An EPA list of hazardous waste of a general nature originating from commercial chemical products, intermediates, or residues.

Uncertainty. The precision of a single measurement, usually the reading of a thermometer or buret or other device that has divisions.

V

Volatile. Readily vaporized. Volatile substances evaporate quickly. Analytes that are volatile are easily lost on being handled and care must be taken to avoid their loss, called volatility losses.

Volatile Organic Compounds (VOCs). Organic compounds of low boiling point which may be easily lost in routine analytical methods. VOCs are analyzed by first separating them from the less volatile components.

W

Wastewater. Used water that contains dissolved and/or suspended matter. It includes mine drainage, landfill leachate, industrial effluents, as well as water discharged from residences.

Weathering. When a pollutant enters the environment it undergoes many changes, including dissolution, vaporization, photolysis, and hydrolysis. Altogether, these processes constitute weathering. A fuel oil, on weathering, becomes enriched in the less-volatile components.

Wetting Agent. A substance that reduces the surface tension of water, allowing it to solubilize dirt and oil, washing them away.

Wet Weight. The mass of a sample that has not been dried. These masses are not as reproducible as dry weights.

Wind Rose. A diagram that summarizes statistical information about wind directions. It is a plot of the direction the wind blows over time.

X

Xenobiotic. A substance not naturally found in biological systems. DDT and its metabolites are xenobiotics in animals.

ACRONYMS

AAQS	Ambient Air Quality Standards
ANSI	American National Standards Institute
APHA	American Public Health Agency
BAT	Best Available Technology
CAA	Clean Air Act
CERCLA	Comprehensive Environmental Response, Compensation, and Liability Act (also known as the Superfund)
CFR	Code of Federal Regulations
CHO	Chemical Hygiene Officer
CHP	Chemical Hygiene Plan
CWA	Clean Water Act
DEP	Department of Environmental Pollution
DER	Department of Environmental Regulation
DOT	Department of Transportation
EPA	U.S. Environmental Protection Agency
FDA	U.S. Food and Drug Administration

HAPs	Hazardous Air Pollutants
HDPE	High-Density Polyethylene
HMIS	Hazardous Material Identification System
HMTA	Hazardous Material Transportation Act
HRS	Health and Rehabilitation Services
HSWA	Hazardous and Solid Waste Act
IDL	Instrument Detection Limit
ISE	Ion-Selective Electrode
LQG	Large-Quantity Generator (of hazardous wastes)
LRB	Laboratory Reagent Blank
MDL	Method Detection Limit
MSDS	Material Safety Data Sheet
NFPA	National Fire Protection Agency
NIOSH	National Institute for Occupational Safety and Health
NIST	National Institute for Standards and Technology (formerly NBS)
NPDES	National Pollutants Discharge Elimination System
NPL	National Priority List
NRC	National Regulatory Commission

OES	Occupational Exposure Standards
OSHA	U.S. Occupational Safety and Health Administration
PCDFs	Polychlorinated Dibenzofurans
PEL	Permissible Exposure Limit (from OSHA)
QA	Quality Assurance
QC	Quality Control
RCRA	Resource Conservation and Recovery Act (of 1976)
SARA	Superfund Amendments and Reauthorization Act
SOP	Standard Operating Procedure
SQG	Small-Quantity Generator (of hazardous waste)
TCDD	2,3,7,8-Tetrachlorodibenzo-*p*-dioxin
TCLP	Toxicity Characteristic Leaching Procedure
TS	Total Solids
TSCA	Toxic Substances Control Act
TSS	Total Suspended Solids
USGS	U.S. Geological Survey

VOC	Volatile Organic Compound
WHMIS	Workplace Hazardous Materials Information System

INDEX

A

B

D

E

F

G

H

I

K

L

N

O

P

Q

R

S

X

Z